Eureka Math
Grade 3
Modules 4 & 5

Special thanks go to the Gordon A. Cain Center and to the Department of Mathematics at Louisiana State University for their support in the development of *Eureka Math*.

For a free *Eureka Math* Teacher Resource Pack, Parent Tip Sheets, and more please visit www.Eureka.tools

Name _____   Date _____

1. Use triangle pattern blocks to cover each shape below. Draw lines to show where the triangles meet. Then, write how many triangle pattern blocks it takes to cover each shape.

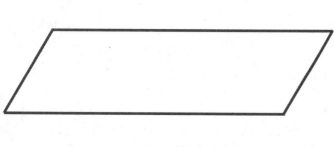

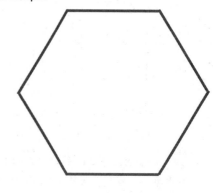

Shape A: _____ triangles

Shape B: _____ triangles

2. Use rhombus pattern blocks to cover each shape below. Draw lines to show where the rhombuses meet. Then, write how many rhombus pattern blocks it takes to cover each shape.

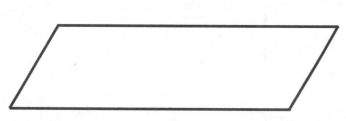

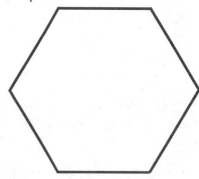

Shape A: _____ rhombuses

Shape B: _____ rhombuses

3. Use trapezoid pattern blocks to cover each shape below. Draw lines to show where the trapezoids meet. Then, write how many trapezoid pattern blocks it requires to cover each shape.

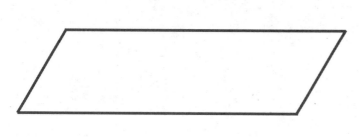

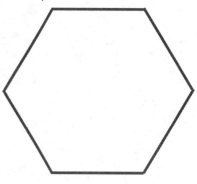

Shape A: _____ trapezoids

Shape B: _____ trapezoids

EUREKA MATH™

Lesson 1:   Understand area as an attribute of plane figures.

1

©2015 Great Minds. eureka-math.org
G3-M4-SE-B3-1.3.1-01.2016

4. How is the number of pattern blocks needed to cover the same shape related to the size of the pattern blocks?

5. Use square pattern blocks to cover the rectangle below. Draw lines to show where the squares meet. Then, write how many square pattern blocks it requires to cover the rectangle.

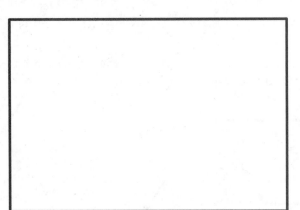

_____ squares

6. Use trapezoid pattern blocks to cover the rectangle in Problem 5. Can you use trapezoid pattern blocks to measure the area of this rectangle? Explain your answer.

Lesson 1:   Understand area as an attribute of plane figures.

EUREKA
MATH™

Name _____ Date _____

1. Magnus covers the same shape with triangles, rhombuses, and trapezoids.

   a. How many triangles will it take to cover the shape?

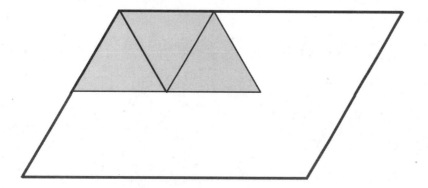

   _____ triangles

   b. How many rhombuses will it take to cover the shape?

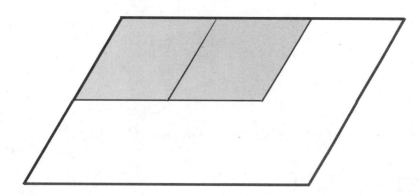

   _____ rhombuses

   c. Magnus notices that 3 triangles from Part (a) cover 1 trapezoid. How many trapezoids will you need to cover the shape below? Explain your answer.

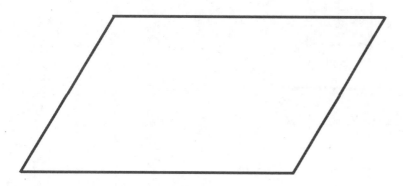

   _____ trapezoids

2.   Angela uses squares to find the area of a rectangle. Her work is shown below.

    a.   How many squares did she use to cover the rectangle?

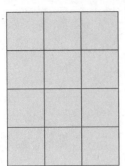   _____ squares

    b.   What is the area of the rectangle in square units? Explain how you found your answer.

3.   Each 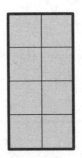 is 1 square unit. Which rectangle has the largest area? How do you know?

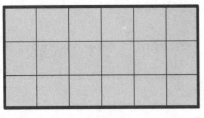

       Rectangle A

                     Rectangle B

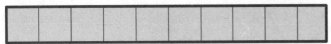

       Rectangle C

EUREKA MATH™

Name _____ Date _____

1. Use all of Paper Strip 1, which you cut into 12 square inches, to complete the chart below.

	Drawing	Area
Rectangle A		
Rectangle B		
Rectangle C		

2. Use all of Paper Strip 2, which you cut into 12 square centimeters, to complete the chart below.

	Drawing	Area
Rectangle A		
Rectangle B		
Rectangle C		

©2015 Great Minds. eureka-math.org
G3-M4-SE-B3-1.3.1-01.2016

3. Compare the areas of the rectangles you made with Paper Strip 1 and Paper Strip 2. What changed? Why did it change?

4. Maggie uses square units to create these two rectangles. Do the two rectangles have the same area? How do you know?

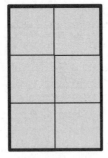

Shape A

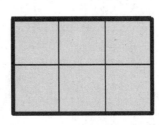

Shape B

5. Count to find the area of the rectangle below. Then, draw a different rectangle that has the same area.

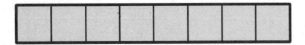

Lesson 2: Decompose and recompose shapes to compare areas.

EUREKA MATH

©2015 Great Minds. eureka-math.org
G3-M4-SE-B3-1.3.1-01.2016

Name _____   Date _____

1.   Each ⬜ is a square unit.   Count to find the area of each rectangle.   Then, circle all the rectangles with an area of 12 square units.

a.

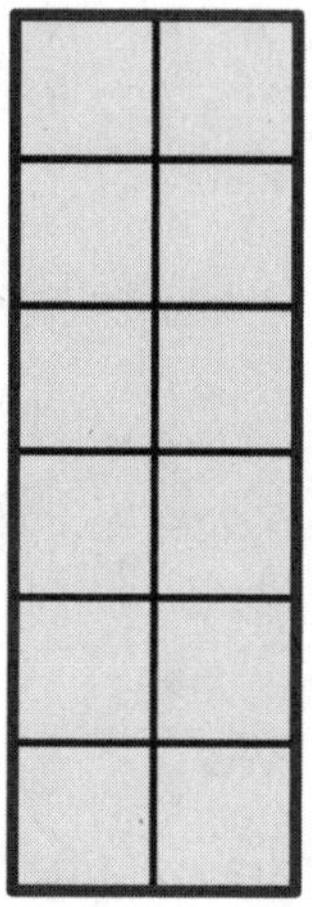

Area = _____ square units

b.

Area = _____ square units

c.

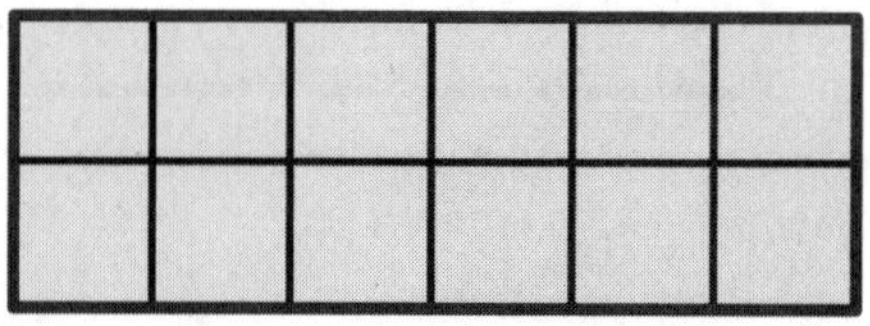

Area = _____ square units

d.

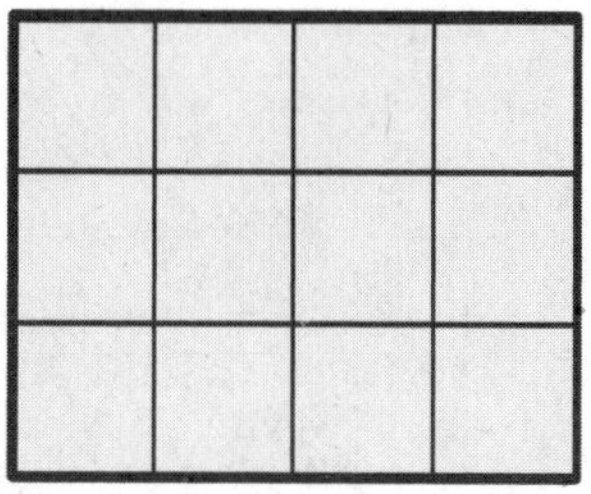

Area = _____ square units

e.

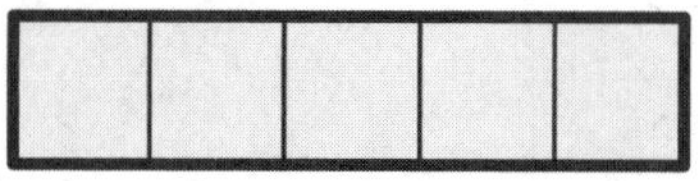

Area = _____ square units

f.

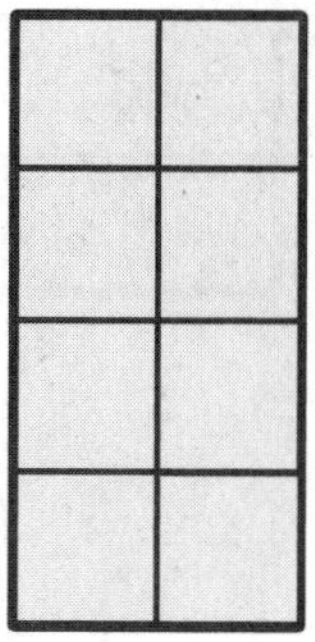

Area = _____ square units

EUREKA MATH™

2. Colin uses square units to create these rectangles. Do they have the same area? Explain.

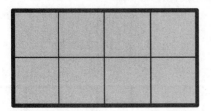

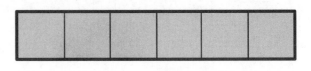

3. Each 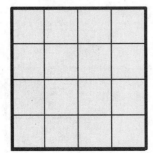 is a square unit. Count to find the area of the rectangle below. Then, draw a different rectangle that has the same area.

Lesson 2:   Decompose and recompose shapes to compare areas.

EUREKA
MATH™

Name _____ Date _____

1. Each ☐ is 1 square unit. What is the area of each of the following rectangles?

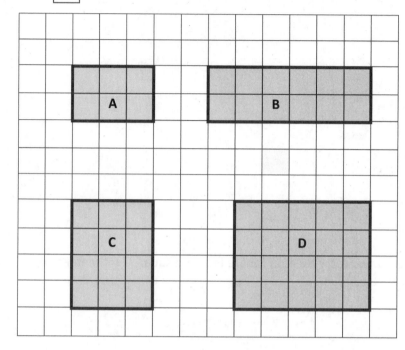

A: _____ square units

B: _____

C: _____

D: _____

2. Each ☐ is 1 square unit. What is the area of each of the following rectangles?

a.

_____

b.

_____

c.

_____

d.

_____

EUREKA MATH

Lesson 3:   Model tiling with centimeter and inch unit squares as a strategy to measure area.

©2015 Great Minds. eureka-math.org
G3-M4-SE-B3-1.3.1-01.2016

9

3.   a.   How would the rectangles in Problem 1 be different if they were composed of square inches?

     b.   Select one rectangle from Problem 1 and recreate it on square inch and square centimeter grid paper.

4.   Use a separate piece of square centimeter grid paper. Draw four different rectangles that each has an area of 8 square centimeters.

Lesson 3:   Model tiling with centimeter and inch unit squares as a strategy to measure area.

©2015 Great Minds. eureka-math.org
G3-M4-SE-B3-1.3.1-01.2016

Name _____ Date _____

1. Each ☐ is 1 square unit. What is the area of each of the following rectangles?

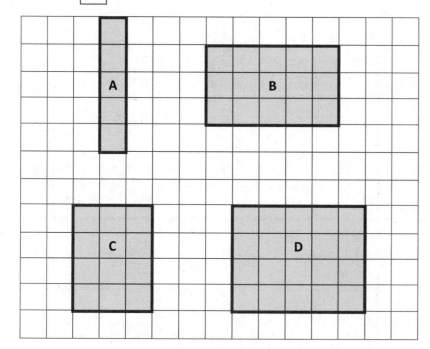

A: _____ square units

B: _____

C: _____

D: _____

2. Each ☐ is 1 square unit. What is the area of each of the following rectangles?

a.

_____

b.

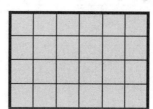

_____

c.

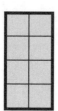

_____

d.

_____

EUREKA
MATH™

Lesson 3:   Model tiling with centimeter and inch unit squares as a strategy to
             measure area.

©2015 Great Minds. eureka-math.org
G3-M4-SE-B3-1.3.1-01.2016

11

3. Each ☐ is 1 square unit. Write the area of each rectangle. Then, draw a different rectangle with the same area in the space provided.

Area = _____ square units

Area = _____

Area = _____

Lesson 3: Model tiling with centimeter and inch unit squares as a strategy to measure area.

©2015 Great Minds. eureka-math.org
G3-M4-SE-B3-1.3.1-01.2016

EUREKA MATH

centimeter grid

Lesson 3:   Model tiling with centimeter and inch unit squares as a strategy to measure area.

©2015 Great Minds. eureka-math.org
G3-M4-SE-B3-1.3.1-01.2016

13

This page intentionally left blank

inch grid

Lesson 3:   Model tiling with centimeter and inch unit squares as a strategy to
             measure area.

©2015 Great Minds. eureka-math.org
G3-M4-SE-B3-1.3.1-01.2016

15

This page intentionally left blank

Name _____ Date _____

1. Use a ruler to measure the side lengths of the rectangle in centimeters. Mark each centimeter with a point and connect the points to show the square units. Then, count the squares you drew to find the total area.

   Total area: _____

2. Use a ruler to measure the side lengths of the rectangle in inches. Mark each inch with a point and connect the points to show the square units. Then, count the squares you drew to find the total area.

   Total area: _____

3. Mariana uses square centimeter tiles to find the side lengths of the rectangle below. Label each side length. Then, count the tiles to find the total area.

   Total area: _____

Lesson 4: Relate side lengths with the number of tiles on a side.

©2015 Great Minds. eureka-math.org
G3-M4-SE-B3-1.3.1-01.2016

17

4. Each 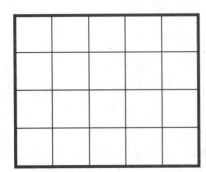 is 1 square centimeter. Saffron says that the side length of the rectangle below is 4 centimeters. Kevin says the side length is 5 centimeters. Who is correct? Explain how you know.

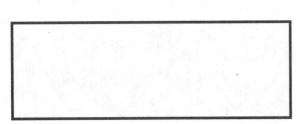

5. Use both square centimeter and square inch tiles to find the area of the rectangle below. Which works best? Explain why.

6. How does knowing side lengths A and B help you find side lengths C and D on the rectangle below?

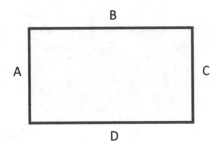

Lesson 4:   Relate side lengths with the number of tiles on a side.

EUREKA
MATH™

Name _____   Date _____

1. Ella placed square centimeter tiles on the rectangle below, and then labeled the side lengths. What is the area of her rectangle?

4 cm

2 cm

Total area: _____

2. Kyle uses square centimeter tiles to find the side lengths of the rectangle below. Label each side length. Then, count the tiles to find the total area.

Total area: _____

3. Maura uses square inch tiles to find the side lengths of the rectangle below. Label each side length. Then, find the total area.

Total area: _____

4. Each square unit below is 1 square inch. Claire says that the side length of the rectangle below is 3 inches. Tyler says the side length is 5 inches. Who is correct? Explain how you know.

5. Label the unknown side lengths for the rectangle below, and then find the area. Explain how you used the lengths provided to find the unknown lengths and area.

4 inches

2 inches

_____

_____

Total area: _____

EUREKA
MATH™

Name _____   Date _____

1. Use the centimeter side of a ruler to draw in the tiles, and then skip-count to find the unknown area. Write a multiplication sentence for each tiled rectangle.

   a. Area: **18** square centimeters.

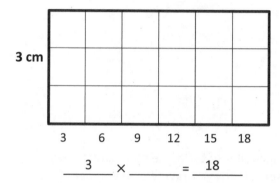

   3 cm

   3   6   9   12   15   18

   ___3___ × _____ = __18__

   d. Area: **24** square centimeters.

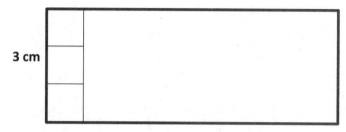

   3 cm

   _____ × _____ = _____

   b. Area: _____ square centimeters.

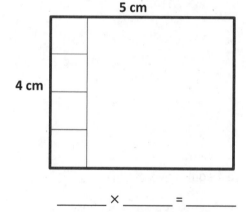

   5 cm

   4 cm

   _____ × _____ = _____

   e. Area: **20** square centimeters.

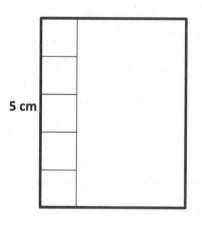

   5 cm

   _____ × _____ = _____

   c. Area: **18** square centimeters.

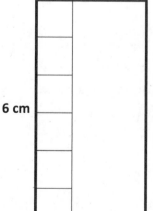

   6 cm

   _____ × _____ = _____

   f. Area: _____ square centimeters.

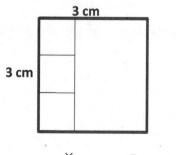

   3 cm

   3 cm

   _____ × _____ = _____

EUREKA MATH™

Lesson 5:   Form rectangles by tiling with unit squares to make arrays.

21

2. Lindsey makes a rectangle with 35 square inch tiles. She arranges the tiles in 5 equal rows. What are the side lengths of the rectangle? Use words, pictures, and numbers to support your answer.

3. Mark has a total of 24 square inch tiles. He uses 18 square inch tiles to build one rectangular array. He uses the remaining square inch tiles to build a second rectangular array. Draw two arrays that Mark might have made. Then, write multiplication sentences for each.

4. Leon makes a rectangle with 32 square centimeter tiles. There are 4 equal rows of tiles.

   a. How many tiles are in each row? Use words, pictures, and numbers to support your answer.

   b. Can Leon arrange all of his 32 square centimeter tiles into 6 equal rows? Explain your answer.

EUREKA
MATH™

Name _____ Date _____

1. Use the centimeter side of a ruler to draw in the tiles, and then skip-count to find the unknown area. Write a multiplication sentence for each tiled rectangle.

a. Area: **24** square centimeters.

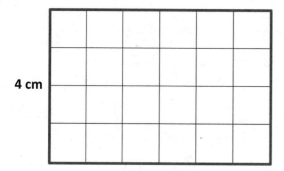

4 cm

___4___ × _____ = ___24___

b. Area: **24** square centimeters.

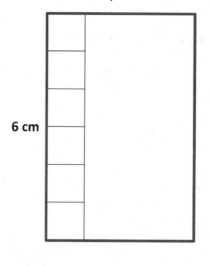

6 cm

_____ × _____ = _____

c. Area: **15** square centimeters.

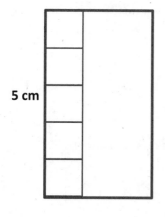

5 cm

_____ × _____ = _____

d. Area: **15** square centimeters.

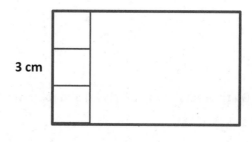

3 cm

_____ × _____ = _____

EUREKA MATH™

Lesson 5: Form rectangles by tiling with unit squares to make arrays.

23

2. Ally makes a rectangle with 45 square inch tiles. She arranges the tiles in 5 equal rows. How many square inch tiles are in each row? Use words, pictures, and numbers to support your answer.

3. Leon makes a rectangle with 36 square centimeter tiles. There are 4 equal rows of tiles.

    a. How many tiles are in each row? Use words, pictures, and numbers to support your answer.

    b. Can Leon arrange all of his 36 square centimeter tiles into 6 equal rows? Use words, pictures, and numbers to support your answer.

    c. Do the rectangles in Parts (a) and (b) have the same total area? Explain how you know.

Lesson 5:   Form rectangles by tiling with unit squares to make arrays.

EUREKA
MATH™

Name _____   Date _____

1. Each ☐ represents 1 square centimeter. Draw to find the number of rows and columns in each array. Match it to its completed array. Then, fill in the blanks to make a true equation to find each array's area.

a. 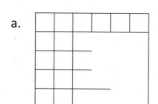   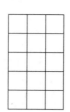

                                      _____ cm × _____ cm = _____ sq cm

b. 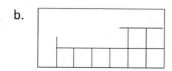   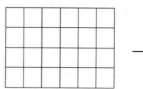

                                      _____ cm × _____ cm = _____ sq cm

c.    

                                      _____ cm × _____ cm = _____ sq cm

d. 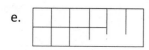   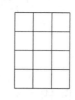

                                      _____ cm × _____ cm = _____ sq cm

e.   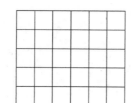

                                      _____ cm × _____ cm = _____ sq cm

f.   _____ cm × _____ cm = _____ sq cm

2. Sheena skip-counts by sixes to find the total square units in the rectangle below. She says there are 42 square units. Is she right? Explain your answer.

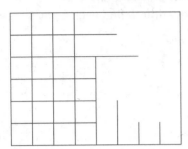

3. The tile floor in Brandon's living room has a rug on it as shown below. How many square tiles are on the floor, including the tiles under the rug?

4. Abdul is creating a stained glass window with square inch glass tiles as shown below. How many more square inch glass tiles does Abdul need to finish his glass window? Explain your answer.

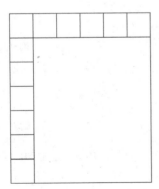

Lesson 6:   Draw rows and columns to determine the area of a rectangle given an incomplete array.

EUREKA
MATH™

Name _____ Date _____

1. Each ▢ represents 1 square centimeter. Draw to find the number of rows and columns in each array. Match it to its completed array. Then, fill in the blanks to make a true equation to find each array's area.

a.

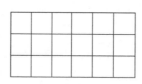

_____ cm × _____ cm = _____ sq cm

b.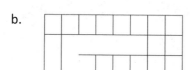

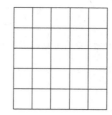

_____ cm × _____ cm = _____ sq cm

c.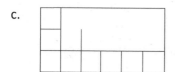

_____ cm × _____ cm = _____ sq cm

_____ cm × _____ cm = _____ sq cm

d.

_____ cm × _____ cm = _____ sq cm

e.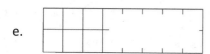

f.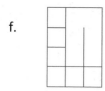

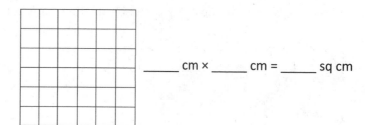

_____ cm × _____ cm = _____ sq cm

EUREKA MATH™

Lesson 6: Draw rows and columns to determine the area of a rectangle given an incomplete array.

©2015 Great Minds. eureka-math.org
G3-M4-SE-B3-1.3.1-01.2016

27

2. Minh skip-counts by sixes to find the total square units in the rectangle below. She says there are 36 square units. Is she correct? Explain your answer.

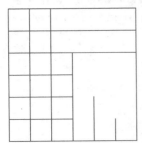

3. The tub in Paige's bathroom covers the tile floor as shown below. How many square tiles are on the floor, including the tiles under the tub?

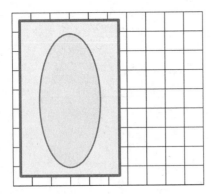

4. Frank sees a book on top of his chessboard. How many squares are covered by the book? Explain your answer.

Lesson 6: Draw rows and columns to determine the area of a rectangle given an incomplete array.

©2015 Great Minds. eureka-math.org
G3-M4-SE-B3-1.3.1-01.2016

EUREKA
MATH™

array 1

Lesson 6:   Draw rows and columns to determine the area of a rectangle given an
             incomplete array.

©2015 Great Minds. eureka-math.org
G3-M4-SE-B3-1.3.1-01.2016

29

This page intentionally left blank

array 2

Lesson 6:   Draw rows and columns to determine the area of a rectangle given an
              incomplete array.

31

©2015 Great Minds. eureka-math.org
G3-M4-SE-B3-1.3.1-01.2016

This page intentionally left blank

Name _____   Date _____

1. Use a straight edge to draw a grid of equal size squares within the rectangle. Find and label the side lengths. Then, multiply the side lengths to find the area.

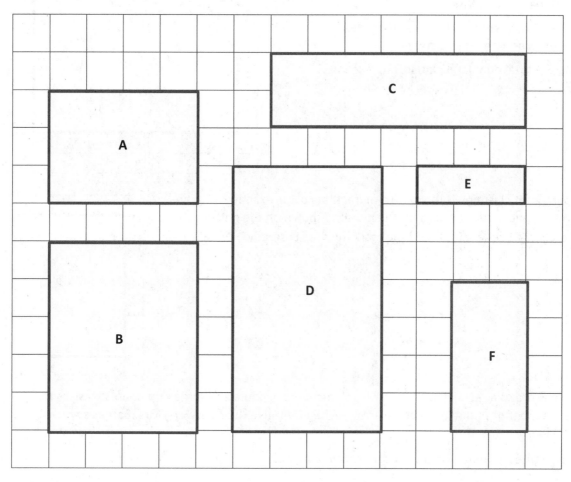

a. Area A:

   ____ units × ____ units = ____ square units

b. Area B:

   ____ units × ____ units = ____ square units

c. Area C:

   ____ units × ____ units = ____ square units

d. Area D:

   ____ units × ____ units = ____ square units

e. Area E:

   ____ unit × ____ units = ____ square units

f. Area F:

   ____ units × ____ units = ____ square units

2. The area of Benjamin's bedroom floor is shown on the grid
    to the right. Each ☐ represents 1 square foot. How
    many total square feet is Benjamin's floor?

    a. Label the side lengths.

    b. Use a straight edge to draw a grid of equal
        size squares within the rectangle.

    c. Find the total number of squares.

Benjamin's
Bedroom Floor

3. Mrs. Young's art class needs to create a mural that covers exactly 35 square
    feet. Mrs. Young marks the area for the mural as shown on the grid. Each
    ☐ represents 1 square foot. Did she mark the area correctly? Explain your
    answer.

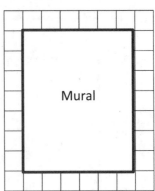

Mural

4. Mrs. Barnes draws a rectangular array. Mila skip-counts by fours and Jorge skip-counts by sixes to find
    the total number of square units in the array. When they give their answers, Mrs. Barnes says that they
    are both right.

    a. Use pictures, numbers, and words to explain how Mila and Jorge can both be right.

    b. How many square units might Mrs. Barnes' array have had?

Lesson 7:   Interpret area models to form rectangular arrays.

EUREKA
MATH™

Name _____   Date _____

1. Find the area of each rectangular array. Label the side lengths of the matching area model, and write a multiplication equation for each area model.

Rectangular Arrays	Area Models
a. _____ square units	3 units 2 units 3 units × _____ units = _____ square units
b. _____ square units	_____ units × _____ units = _____ square units
c. _____ square units	_____ units × _____ units = _____ square units
d. _____ square units	_____ units × _____ units = _____ square units

2. Jillian arranges square pattern blocks into a 7 by 4 array. Draw Jillian's array on the the grid below. How many square units are in Jillian's rectangular array?

a.

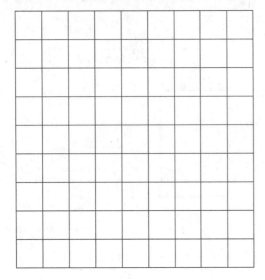

b. Label the side lengths of Jillian's array from Part (a) on the rectangle below. Then, write a multiplication sentence to represent the area of the rectangle.

3. Fiona draws a 24 square centimeter rectangle. Gregory draws a 24 square inch rectangle. Whose rectangle is larger in area? How do you know?

EUREKA
MATH™

area model

This page intentionally left blank

Name _____ Date _____

1. Write a multiplication equation to find the area of each rectangle.

a.
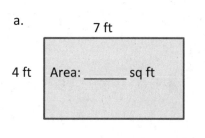
7 ft

4 ft Area: _____ sq ft

_____ × _____ = _____

b.
7 ft

8 ft Area: _____ sq ft

_____ × _____ = _____

c.
6 ft

6 ft Area: _____ sq ft

_____ × _____ = _____

2. Write a multiplication equation and a division equation to find the unknown side length for each rectangle.

a.
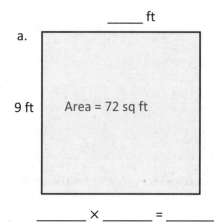
_____ ft

9 ft Area = 72 sq ft

_____ × _____ = _____

_____ ÷ _____ = _____

b.
_____ ft

3 ft Area = 15 sq ft

_____ × _____ = _____

_____ ÷ _____ = _____

c.
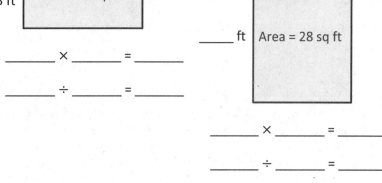
4 ft

_____ ft Area = 28 sq ft

_____ × _____ = _____

_____ ÷ _____ = _____

3. On the grid below, draw a rectangle that has an area of 42 square units. Label the side lengths.

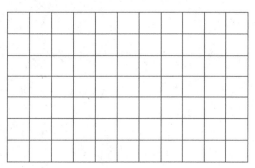

Lesson 8:   Find the area of a rectangle through multiplication of the side lengths.

39

EUREKA
MATH

4. Ursa draws a rectangle that has side lengths of 9 centimeters and 6 centimeters. What is the area of the rectangle? Explain how you found your answer.

5. Eliza's bedroom measures 6 feet by 7 feet. Her brother's bedroom measures 5 feet by 8 feet. Eliza says their rooms have the same exact floor area. Is she right? Why or why not?

6. Cliff draws a rectangle with a side length of 6 inches and an area of 24 square inches. What is the other side length? How do you know?

Lesson 8:   Find the area of a rectangle through multiplication of the side lengths.

EUREKA
MATH™

Name _____   Date _____

1. Write a multiplication equation to find the area of each rectangle.

a.

8 cm

3 cm | Area: _____ sq cm

_____ × _____ = _____

b.

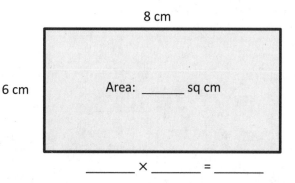

8 cm

6 cm | Area: _____ sq cm

_____ × _____ = _____

c.

4 ft

4 ft | Area: _____ sq ft

_____ × _____ = _____

d.

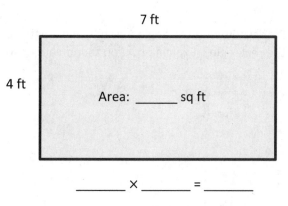

7 ft

4 ft | Area: _____ sq ft

_____ × _____ = _____

2. Write a multiplication equation and a division equation to find the unknown side length for each rectangle.

a.

_____ ft.

3 ft | Area: 24 sq ft

_____ × _____ = _____

_____ ÷ _____ = _____

b.

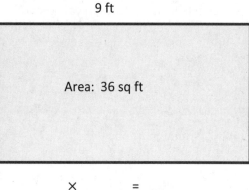

9 ft

_____ ft | Area: 36 sq ft

_____ × _____ = _____

_____ ÷ _____ = _____

Lesson 8:   Find the area of a rectangle through multiplication of the side lengths.

©2015 Great Minds. eureka-math.org
G3-M4-SE-B3-1.3.1-01.2016

41

3. On the grid below, draw a rectangle that has an area of 32 square centimeters. Label the side lengths.

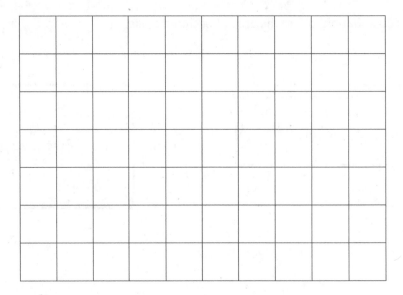

4. Patricia draws a rectangle that has side lengths of 4 centimeters and 9 centimeters. What is the area of the rectangle? Explain how you found your answer.

5. Charles draws a rectangle with a side length of 9 inches and an area of 27 square inches. What is the other side length? How do you know?

      Lesson 8:   Find the area of a rectangle through multiplication of the side lengths.

EUREKA
MATH™

grid

Lesson 8:   Find the area of a rectangle through multiplication of the side lengths.

43

©2015 Great Minds. eureka-math.org
G3-M4-SE-B3-1.3.1-01.2016

This page intentionally left blank

Name _____   Date _____

1. Cut the grid into 2 equal rectangles.

   a. Draw and label the side lengths of the 2 rectangles.

   b. Write an equation to find the area of 1 of the rectangles.

   c. Write an equation to show the total area of the 2 rectangles.

2. Place your 2 equal rectangles side by side to create a new, longer rectangle.

   a. Draw an area model to show the new rectangle. Label the side lengths.

   b. Find the total area of the longer rectangle.

3. Furaha and Rahema use square tiles to make the rectangles shown below.

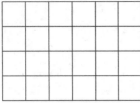

Furaha's Rectangle

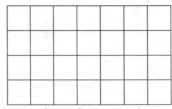

Rahema's Rectangle

a. Label the side lengths on the rectangles above, and find the area of each rectangle.

b. Furaha pushes his rectangle next to Rahema's rectangle to form a new, longer rectangle. Draw an area model to show the new rectangle. Label the side lengths.

c. Rahema says the area of the new, longer rectangle is 52 square units. Is she right? Explain your answer.

4. Kiera says she can find the area of the long rectangle below by adding the areas of Rectangles A and B. Is she right? Why or why not?

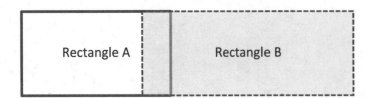

Rectangle A   Rectangle B

Lesson 9:   Analyze different rectangles and reason about their area.

Name _____   Date _____

1. Use the grid to answer the questions below.

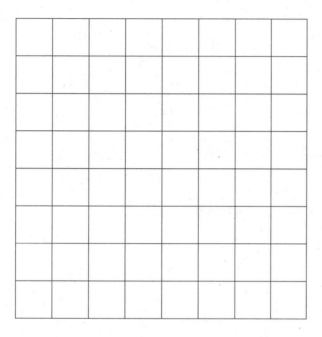

a. Draw a line to divide the grid into 2 equal rectangles. Shade in 1 of the rectangles that you created.

b. Label the side lengths of each rectangle.

c. Write an equation to show the total area of the 2 rectangles.

2. Alexa cuts out the 2 equal rectangles from Problem 1(a) and puts the two shorter sides together.

   a. Draw Alexa's new rectangle and label the side lengths below.

   b. Find the total area of the new, longer rectangle.

   c. Is the area of the new, longer rectangle equal to the total area in Problem 1(c)?
      Explain why or why not.

Lesson 9:   Analyze different rectangles and reason about their area.

Name _____   Date _____

1. Label the side lengths of the shaded and unshaded rectangles when needed. Then, find the total area of the large rectangle by adding the areas of the two smaller rectangles.

a.

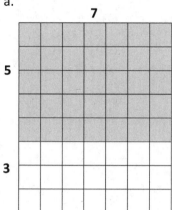

$8 \times 7 = (5 + 3) \times 7$

     $= (5 \times 7) + (3 \times 7)$

     $=$ _____ + _____

     $=$ _____

Area: _____ square units

b.

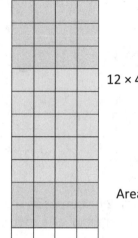

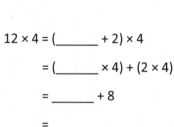

$12 \times 4 = ($ _____ $+ 2) \times 4$

     $= ($ _____ $\times 4) + (2 \times 4)$

     $=$ _____ $+ 8$

     $=$ _____

Area: _____ square units

c.

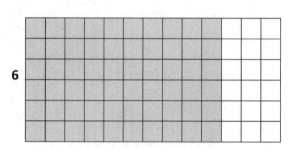

$6 \times 13 = 6 \times ($ _____ $+ 3)$

     $= (6 \times$ _____ $) + (6 \times 3)$

     $=$ _____ $+$ _____

     $=$ _____

Area: _____ square units

d.

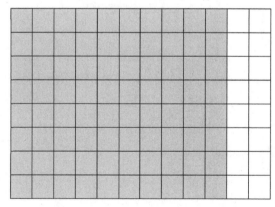

$8 \times 12 = 8 \times ($ _____ $+$ _____ $)$

     $= (8 \times$ _____ $) + (8 \times$ _____ $)$

     $=$ _____ $+$ _____

     $=$ _____

Area: _____ square units

Lesson 10:   Apply the distributive property as a strategy to find the total area of a large rectangle by adding two products.

49

©2015 Great Minds. eureka-math.org
G3-M4-SE-B3-1.3.1-01.2016

2. Vince imagines 1 more row of eight to find the total area of a 9 × 8 rectangle. Explain how this could help him solve 9 × 8.

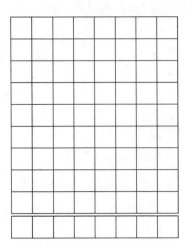

3. Break the 15 × 5 rectangle into 2 rectangles by shading one smaller rectangle within it. Then, find the sum of the areas of the 2 smaller rectangles and show how it relates to the total area. Explain your thinking.

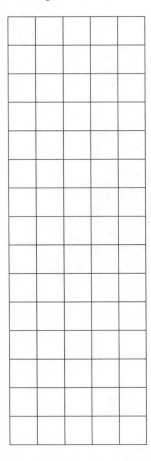

Lesson 10:   Apply the distributive property as a strategy to find the total area of a large rectangle by adding two products.

©2015 Great Minds. eureka-math.org
G3-M4-SE-B3-1.3.1-01.2016

EUREKA
MATH™

Name _____   Date _____

1. Label the side lengths of the shaded and unshaded rectangles. Then, find the total area of the large rectangle by adding the areas of the 2 smaller rectangles.

a.

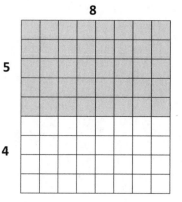

9 × 8 = (5 + 4) × 8

= (5 × 8) + (4 × 8)

= _____ + _____

= _____

Area: _____ square units

b.

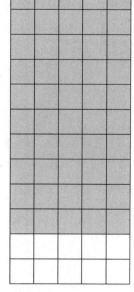

12 × 5 = (_____ + 2) × 5

= (_____ × 5) + (2 × 5)

= _____ + 10

= _____

Area: _____ square units

c.

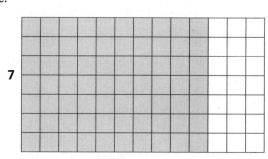

7 × 13 = 7 × (_____ + 3)

= (7 × _____) + (7 × 3)

= _____ + _____

= _____

Area: _____ square units

d.

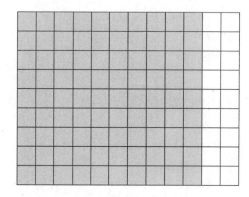

9 × 12 = 9 × (_____ + _____)

= (9 × _____) + (9 × _____)

= _____ + _____

= _____

Area: _____ square units

Lesson 10: Apply the distributive property as a strategy to find the total area of a large rectangle by adding two products.

51

©2015 Great Minds. eureka-math.org
G3-M4-SE-B3-1.3.1-01.2016

2.   Finn imagines 1 more row of nine to find the total area of 9 × 9 rectangle. Explain how this could help him solve 9 × 9.

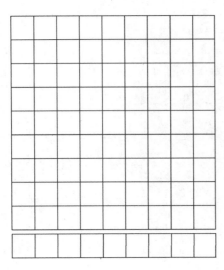

3.   Shade an area to break the 16 × 4 rectangle into 2 smaller rectangles. Then, find the sum of the areas of the 2 smaller rectangles to find the total area. Explain your thinking.

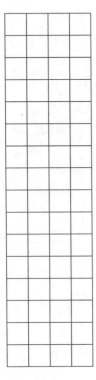

Lesson 10:   Apply the distributive property as a strategy to find the total area of a large rectangle by adding two products.

EUREKA
MATH™

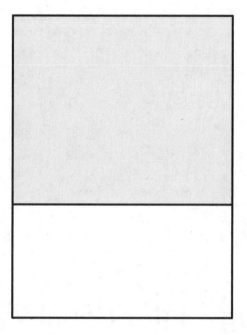

tiling

EUREKA
MATH™

Lesson 10: Apply the distributive property as a strategy to find the total area of a large rectangle by adding two products.

53

©2015 Great Minds. eureka-math.org
G3-M4-SE-B3-1.3.1-01.2016

This page intentionally left blank

Name _____ Date _____

1. The rectangles below have the same area. Move the parentheses to find the unknown side lengths. Then, solve.

a.
6 cm

8 cm

Area: 8 × _____ = _____

Area: _____ sq cm

b.
_____ cm

1 cm

Area: 1 × 48 = _____

Area: _____ sq cm

Area: **8 × 6** = (2 × 4) × 6

= 2 × 4 × 6

= _____ × _____

= _____.

Area: _____ sq cm

c.
_____ cm

2 cm

d.
_____ cm

4 cm

Area: **8 × 6** = (4 × 2) × 6

= 4 × 2 × 6

= _____ × _____

= _____

Area: _____ sq cm

e.
_____ cm

_____ cm

Area: **8 × 6** = 8 × (2 × 3)

= 8 × 2 × 3

= _____ × _____

= _____

Area: _____ sq cm

2. Does Problem 1 show all the possible whole number side lengths for a rectangle with an area of 48 square centimeters? How do you know?

Lesson 11:   Demonstrate the possible whole number side lengths of rectangles with areas of 24, 36, 48, or 72 square units using the associative property.

©2015 Great Minds. eureka-math.org
G3-M4-SE-B3-1.3.1-01.2016

55

EUREKA
MATH™

3. In Problem 1, what happens to the shape of the rectangle as the difference between the side lengths gets smaller?

4. a. Find the area of the rectangle below.

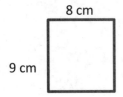

8 cm

9 cm

b. Julius says a 4 cm by 18 cm rectangle has the same area as the rectangle in Part (a). Place parentheses in the equation to find the related fact and solve. Is Julius correct? Why or why not?

   4 × 18 = 4 × 2 × 9

   = 4 × 2 × 9

   = _____ × _____

   = _____

   Area: _____ sq cm

c. Use the expression 8 × 9 to find different side lengths for a rectangle that has the same area as the rectangle in Part (a). Show your equations using parentheses. Then, estimate to draw the rectangle and label the side lengths.

Lesson 11:   Demonstrate the possible whole number side lengths of rectangles with areas of 24, 36, 48, or 72 square units using the associative property.
©2015 Great Minds. eureka-math.org
G3-M4-SE-B3-1.3.1-01.2016

EUREKA
MATH™

Name _____ Date _____

1. The rectangles below have the same area. Move the parentheses to find the unknown side lengths. Then, solve.

36 cm

1 cm

b. Area: 1 × 36 = _____

Area: _____ sq cm

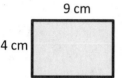

9 cm

4 cm

a. Area: 4 × _____ = _____

Area: _____ sq cm

_____ cm

2 cm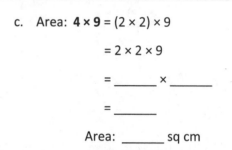

c. Area: **4 × 9** = (2 × 2) × 9

= 2 × 2 × 9

= _____ × _____

= _____

Area: _____ sq cm

_____ cm

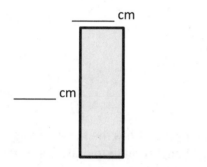

_____ cm

d. Area: **4 × 9** = 4 × (3 × 3)

= 4 × 3 × 3

= _____ × _____

= _____

Area: _____ sq cm

e. Area: **12 × 3** = (6 × 2) × 3

= 6 × 2 × 3

_____ cm

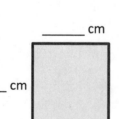

_____ cm

= _____ × _____

= _____

Area: _____ sq cm

2. Does Problem 1 show all the possible whole number side lengths for a rectangle with an area of 36 square centimeters? How do you know?

EUREKA MATH

Lesson 11: Demonstrate the possible whole number side lengths of rectangles with areas of 24, 36, 48, or 72 square units using the associative property.

©2015 Great Minds. eureka-math.org
G3-M4-SE-B3-1.3.1-01.2016

57

3. a. Find the area of the rectangle below.

6 cm

8 cm

b. Hilda says a 4 cm by 12 cm rectangle has the same area as the rectangle in Part (a). Place parentheses in the equation to find the related fact and solve. Is Hilda correct? Why or why not?

4 × 12 = 4 × 2 × 6

= 4 × 2 × 6

= _____ × _____

= _____

Area: _____ sq cm

c. Use the expression 8 × 6 to find different side lengths for a rectangle that has the same area as the rectangle in Part (a). Show your equations using parentheses. Then, estimate to draw the rectangle and label the side lengths.

Lesson 11:   Demonstrate the possible whole number side lengths of rectangles with areas of 24, 36, 48, or 72 square units using the associative property.

©2015 Great Minds. eureka-math.org
G3-M4-SE-B3-1.3.1-01.2016

EUREKA
MATH

Name _____   Date _____

1. Each side on a sticky note measures 9 centimeters. What is the area of the sticky note?

2. Stacy tiles the rectangle below using her square pattern blocks.

a. Find the area of Stacy's rectangle in square units. Then, draw and label a different rectangle with whole number side lengths that has the same area.

b. Can you draw another rectangle with different whole number side lengths and have the same area? Explain how you know.

3. An artist paints a 4 foot × 16 foot mural on a wall. What is the total area of the mural? Use the break apart and distribute strategy.

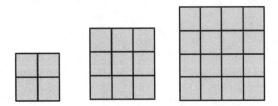

4. Alana tiles the 3 figures below. She says, "I'm making a pattern!"

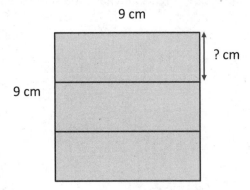

    a. Find the area of Alana's 3 figures and explain her pattern.

    b. Draw the next 2 figures in Alana's pattern and find their areas.

5. Jermaine glues 3 identical pieces of paper as shown below and makes a square. Find the unknown side length of 1 piece of paper. Then, find the total area of 2 pieces of paper.

EUREKA MATH

Name _____ Date _____

1. A square calendar has sides that are 9 inches long. What is the calendar's area?

2. Each [ ] is 1 square unit. Sienna uses the same square units to draw a 6 × 2 rectangle and says that it has the same area as the rectangle below. Is she correct? Explain why or why not.

3. The surface of an office desk has an area of 15 square feet. Its length is 5 feet. How wide is the office desk?

4. A rectangular garden has a total area of 48 square yards. Draw and label two possible rectangular gardens with different side lengths that have the same area.

5. Lila makes the pattern below. Find and explain her pattern. Then, draw the *fifth* figure in her pattern.

Lesson 12:   Solve word problems involving area.

©2015 Great Minds. eureka-math.org
G3-M4-SE-B3-1.3.1-01.2016

EUREKA
MATH™

Name _____   Date _____

1.   Each of the following figures is made up of 2 rectangles. Find the total area of each figure.

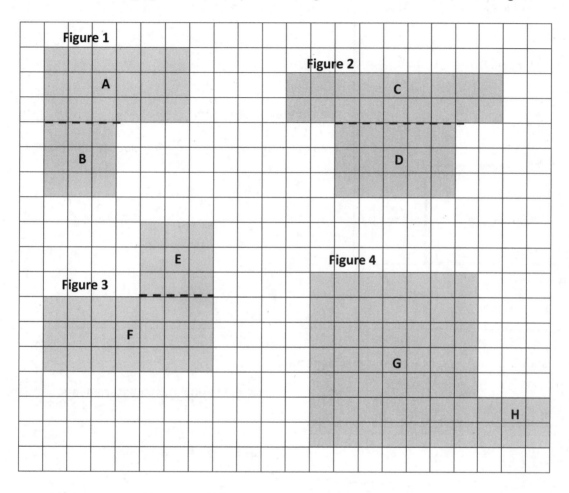

Figure 1: Area of A + Area of B: ___18___ sq units + _____ sq units = _____ sq units

Figure 2: Area of C + Area of D: _____ sq units + _____ sq units = _____ sq units

Figure 3: Area of E + Area of F: _____ sq units + _____ sq units = _____ sq units

Figure 4: Area of G + Area of H: _____ sq units + _____ sq units = _____sq units

Lesson 13:   Find areas by decomposing into rectangles or completing composite
              figures to form rectangles.

©2015 Great Minds. eureka-math.org
G3-M4-SE-B3-1.3.1-01.2016

63

2. The figure shows a small rectangle cut out of a bigger rectangle. Find the area of the shaded figure.

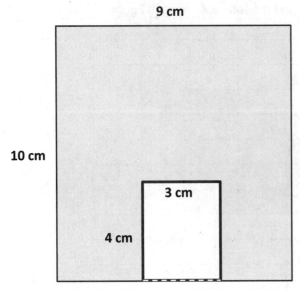

Area of the shaded figure: _____ − _____ = _____

Area of the shaded figure: _____ square centimeters

3. The figure shows a small rectangle cut out of a big rectangle.

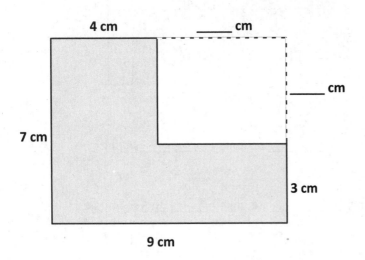

a. Label the unknown measurements.

b. Area of the big rectangle:

   _____ cm × _____ cm = _____ sq cm

c. Area of the small rectangle:

   _____ cm × _____ cm = _____ sq cm

d. Find the area of the shaded figure.

Lesson 13:   Find areas by decomposing into rectangles or completing composite figures to form rectangles.

EUREKA MATH™

Name _____   Date _____

1. Each of the following figures is made up of 2 rectangles. Find the total area of each figure.

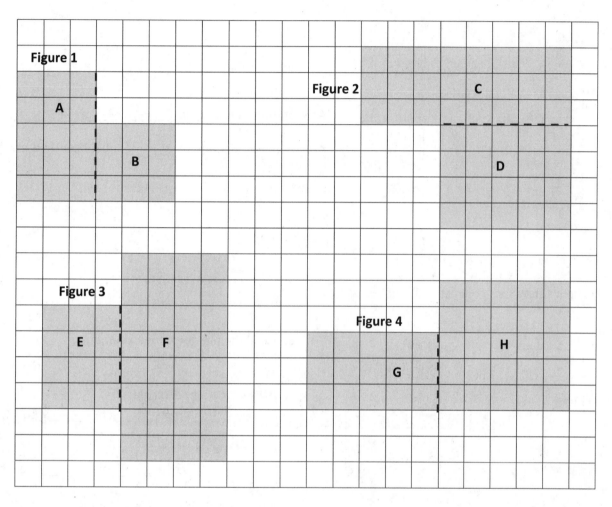

Figure 1: Area of A + Area of B: _____ sq units + _____ sq units = _____ sq units

Figure 2: Area of C + Area of D: _____ sq units + _____ sq units = _____ sq units

Figure 3: Area of E + Area of F: _____ sq units + _____ sq units = _____ sq units

Figure 4: Area of G + Area of H: _____ sq units + _____ sq units = _____ sq units

EUREKA
MATH™

Lesson 13:   Find areas by decomposing into rectangles or completing composite
figures to form rectangles.

©2015 Great Minds. eureka-math.org
G3-M4-SE-B3-1.3.1-01.2016

65

2. The figure shows a small rectangle cut out of a big rectangle. Find the area of the shaded figure.

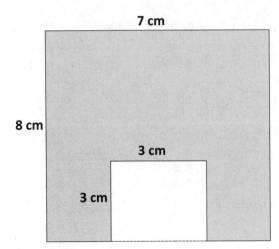

Area of the shaded figure: _____ – _____ = _____

Area of the shaded figure: _____ square centimeters

3. The figure shows a small rectangle cut out of a big rectangle.

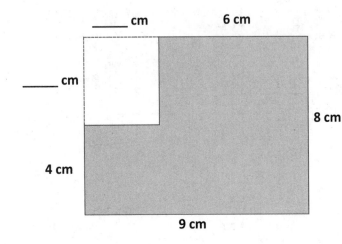

a. Label the unknown measurements.

b. Area of the big rectangle:

    _____ cm × _____ cm = _____ sq cm

c. Area of the small rectangle:

    _____ cm × _____ cm = _____ sq cm

d. Find the area of the shaded figure.

Lesson 13:   Find areas by decomposing into rectangles or completing composite figures to form rectangles.

©2015 Great Minds. eureka-math.org
G3-M4-SE-B3-1.3.1-01.2016

EUREKA MATH™

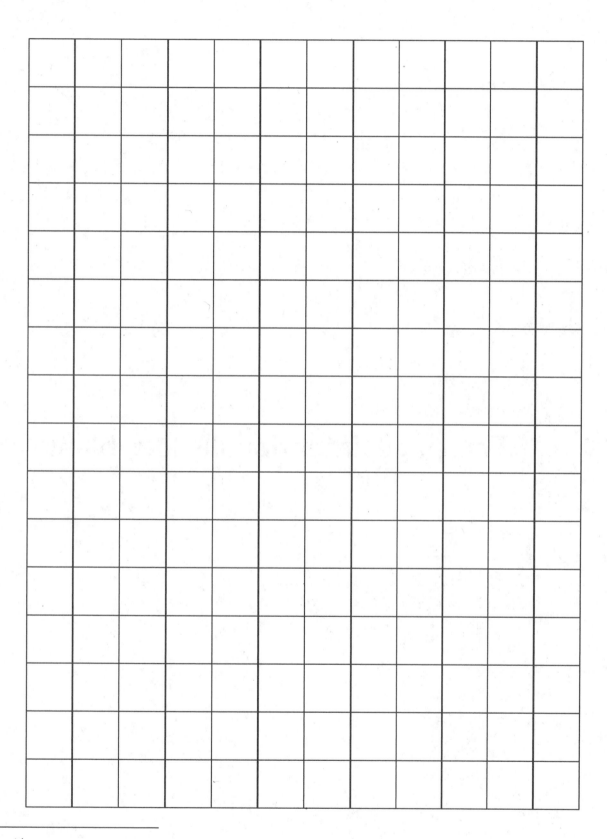

large grid

Lesson 13: Find areas by decomposing into rectangles or completing composite figures to form rectangles.

©2015 Great Minds. eureka-math.org
G3-M4-SE-B3-1.3.1-01.2016

67

This page intentionally left blank

Name _____   Date _____

1. Find the area of each of the following figures. All figures are made up of rectangles.

a.

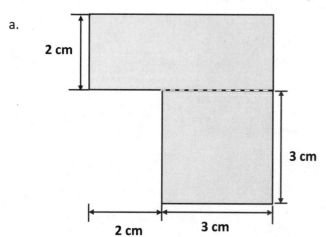

b.

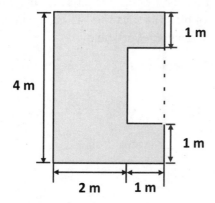

2. The figure below shows a small rectangle in a big rectangle. Find the area of the shaded part of the figure.

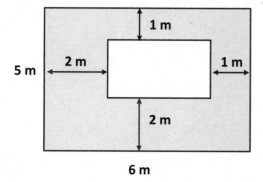

EUREKA
MATH™

Lesson 14:   Find areas by decomposing into rectangles or completing composite figures to form rectangles.

©2015 Great Minds. eureka-math.org
G3-M4-SE-B3-1.3.1-01.2016

69

3. A paper rectangle has a length of 6 inches and a width of 8 inches. A square with a side length of 3 inches was cut out of it. What is the area of the remaining paper?

4. Tila and Evan both have paper rectangles measuring 6 cm by 9 cm. Tila cuts a 3 cm by 4 cm rectangle out of hers, and Evan cuts a 2 cm by 6 cm rectangle out of his. Tila says she has more paper left over. Evan says they have the same amount. Who is correct? Show your work below.

Lesson 14:   Find areas by decomposing into rectangles or completing composite
                            figures to form rectangles.

EUREKA
MATH™

Name _____ Date _____

1. Find the area of each of the following figures. All figures are made up of rectangles.

    a.

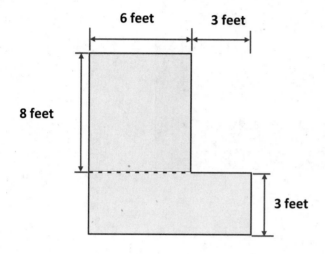

    b.

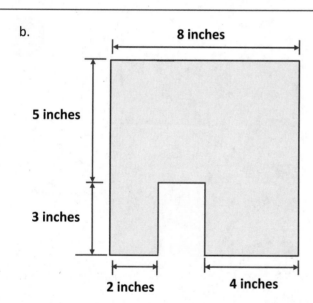

Lesson 14:   Find areas by decomposing into rectangles or completing composite
            figures to form rectangles.

©2015 Great Minds. eureka-math.org
G3-M4-SE-B3-1.3.1-01.2016

71

2. The figure below shows a small rectangle cut out of a big rectangle.

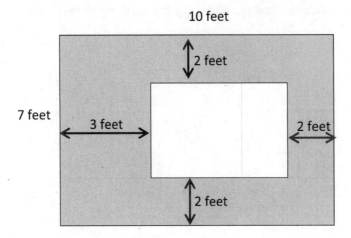

a. Label the side lengths of the unshaded region.

b. Find the area of the shaded region.

Lesson 14: Find areas by decomposing into rectangles or completing composite figures to form rectangles.

©2015 Great Minds. eureka-math.org
G3-M4-SE-B3-1.3.1-01.2016

EUREKA MATH™

Name _____   Date _____

1. Make a prediction: Which room looks like it has the biggest area?

2. Record the areas and show the strategy you used to find each area.

Room	Area	Strategy
Bedroom 1	_____ sq cm	
Bedroom 2	_____ sq cm	
Kitchen	_____ sq cm	
Hallway	_____ sq cm	
Bathroom	_____ sq cm	
Dining Room	_____ sq cm	
Living Room	_____ sq cm	

Lesson 15:   Apply knowledge of area to determine areas of rooms in a given floor
              plan.

©2015 Great Minds. eureka-math.org
G3-M4-SE-B3-1.3.1-01.2016

73

3. Which room has the biggest area? Was your prediction right? Why or why not?

4. Find the side lengths of the house without using your ruler to measure them, and explain the process you used.

   Side lengths: _____ centimeters and _____ centimeters

5. What is the area of the whole floor plan? How do you know?

   Area = _____ square centimeters

Lesson 15: Apply knowledge of area to determine areas of rooms in a given floor plan.

EUREKA
MATH™

The rooms in the floor plan below are rectangles or made up of rectangles.

Bedroom 1		Bathroom
Kitchen	**Hallway**	**Bedroom 2**
Dining Room		
	Living Room	

EUREKA MATH

Lesson 15:   Apply knowledge of area to determine areas of rooms in a given floor plan.

©2015 Great Minds. eureka-math.org
G3-M4-SE-B3-1.3.1-01.2016

75

This page intentionally left blank

Name _____   Date _____

Use a ruler to measure the side lengths of each numbered room in centimeters. Then, find the area. Use the measurements below to match, and label the rooms with the correct areas.

Kitchen: 45 square centimeters   Living Room: 63 square centimeters

Porch: 34 square centimeters   Bedroom: 56 square centimeters

Bathroom: 24 square centimeters   Hallway: 12 square centimeters

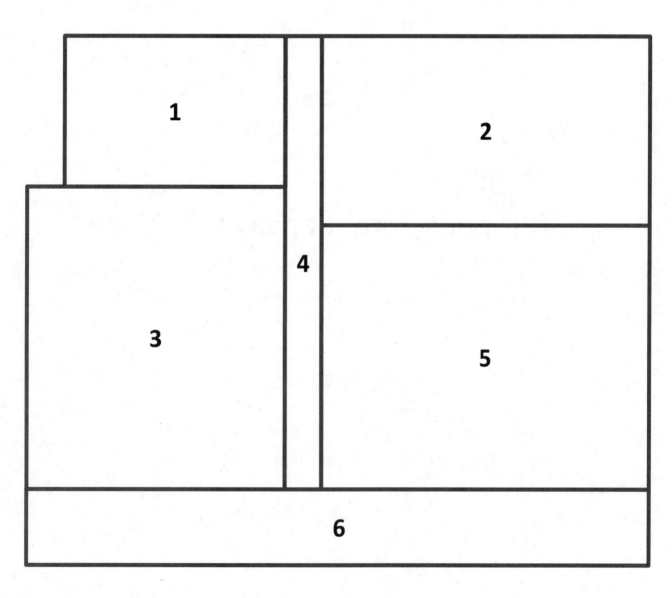

Lesson 15:   Apply knowledge of area to determine areas of rooms in a given floor plan.

©2015 Great Minds. eureka-math.org
G3-M4-SE-B3-1.3.1-01.2016

77

This page intentionally left blank

Name _____ Date _____

Record the new side lengths you have chosen for each of the rooms and show that these side lengths equal the required area. For non-rectangular rooms, record the side lengths and areas of the small rectangles. Then, show how the areas of the small rectangles equal the required area.

Room	New Side Lengths
Bedroom 1: 60 sq cm	
Bedroom 2: 56 sq cm	
Kitchen: 42 sq cm	

Lesson 16:   Apply knowledge of area to determine areas of rooms in a given floor plan.

©2015 Great Minds. eureka-math.org
G3-M4-SE-B3-1.3.1-01.2016

79

Room	New Side Lengths
Hallway: 24 sq cm	
Bathroom: 25 sq cm	
Dining Room: 28 sq cm	
Living Room: 88 sq cm	

Lesson 16:   Apply knowledge of area to determine areas of rooms in a given floor plan.

©2015 Great Minds. eureka-math.org
G3-M4-SE-B3-1.3.1-01.2016

EUREKA
MATH™

Name _____   Date _____

Jeremy plans and designs his own dream playground on grid paper. His new playground will cover a total area of 100 square units. The chart shows how much space he gives for each piece of equipment, or area. Use the information in the chart to draw and label a possible way Jeremy can plan his playground.

Basketball court	10 square units
Jungle gym	9 square units
Slide	6 square units
Soccer area	24 square units

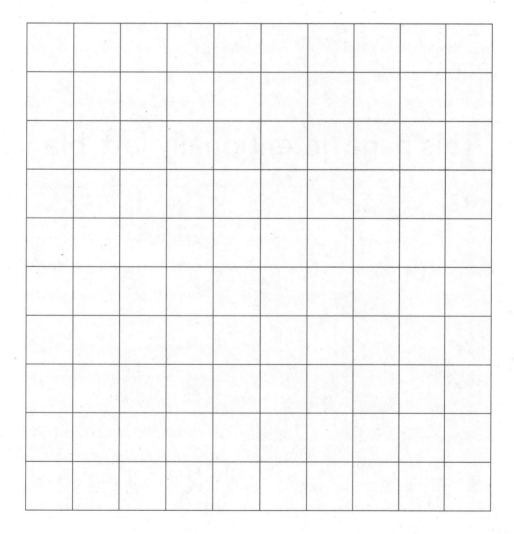

Lesson 16:   Apply knowledge of area to determine areas of rooms in a given floor plan.

©2015 Great Minds. eureka-math.org
G3-M4-SE-B3-1.3.1-01.2016

81

This page intentionally left blank

Eureka Math
Grade 3
Module 5

Special thanks go to the Gordon A. Cain Center and to the Department of Mathematics at Louisiana State University for their support in the development of *Eureka Math*.

For a free *Eureka Math* Teacher Resource Pack, Parent Tip Sheets, and more please visit www.Eureka.tools

Printed in the U.S.A.

This book may be purchased from the publisher at eureka-math.org

1 2 3 4 5 6 7 8   BAB   25 24 23 22 21

ISBN 978-1-63255-300-3

Name _____   Date _____

1. A beaker is considered full when the liquid reaches the fill line shown near the top. Estimate the amount
    of water in the beaker by shading the drawing as indicated. The first one is done for you.

| 1 half | 1 fourth | 1 third |

2. Juanita cut her string cheese into equal pieces as shown in the rectangles below. In the blanks below,
    name the fraction of the string cheese represented by the shaded part.

   _____

   _____

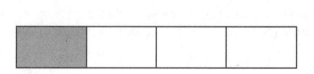   _____

EUREKA MATH™   **Lesson 1:**   Specify and partition a whole into equal parts, identifying and   1
                                   counting unit fractions using concrete models.

©2015 Great Minds. eureka-math.org
G3-M5-SE-B3-1.3.1-01.2016

3. a. In the space below, draw a small rectangle. Estimate to split it into 2 equal parts. How many lines did you draw to make 2 equal parts? What is the name of each fractional unit?

    b. Draw another small rectangle. Estimate to split it into 3 equal parts. How many lines did you draw to make 3 equal parts? What is the name of each fractional unit?

    c. Draw another small rectangle. Estimate to split it into 4 equal parts. How many lines did you draw to make 4 equal parts? What is the name of each fractional unit?

4. Each rectangle represents 1 sheet of paper.

    a. Estimate to show how you would cut the paper into fractional units as indicated below.

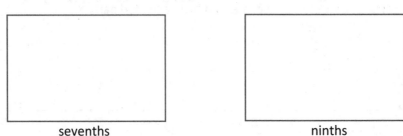

            sevenths   ninths

    b. What do you notice? How many lines do you think you would draw to make a rectangle with 20 equal parts?

5. Rochelle has a strip of wood 12 inches long. She cuts it into pieces that are each 6 inches in length. What fraction of the wood is one piece? Use your strip from the lesson to help you. Draw a picture to show the piece of wood and how Rochelle cut it.

Lesson 1:   Specify and partition a whole into equal parts, identifying and
                         counting unit fractions using concrete models.

EUREKA
MATH™

©2015 Great Minds. eureka-math.org
G3-M5-SE-B3-1.3.1-01.2016

Name _____   Date _____

1. A beaker is considered full when the liquid reaches the fill line shown near the top. Estimate the amount of water in the beaker by shading the drawing as indicated. The first one is done for you.

   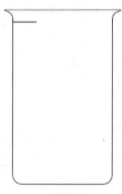   

    1 half   1 fifth   1 sixth

2. Danielle cut her candy bar into equal pieces as shown in the rectangles below. In the blanks below, name the fraction of candy bar represented by the shaded part.

_____   _____   _____

3. Each circle represents 1 whole pie. Estimate to show how you would cut the pie into fractional units as indicated below.

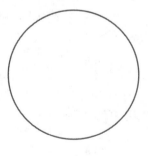   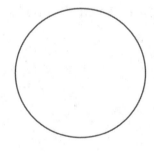   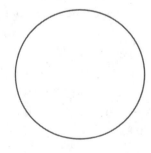

| halves | | thirds | | sixths |

EUREKA MATH

Lesson 1: Specify and partition a whole into equal parts, identifying and counting unit fractions using concrete models.

3

©2015 Great Minds. eureka-math.org
G3-M5-SE-B3-1.3.1-01.2016

4. Each rectangle represents 1 sheet of paper. Estimate to draw lines to show how you would cut the paper into fractional units as indicated below.

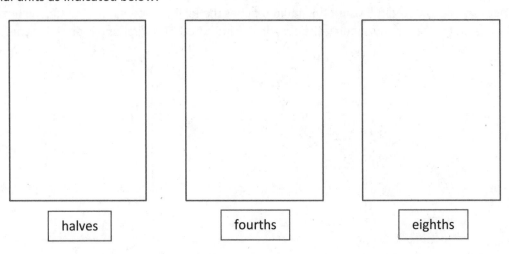

halves   fourths   eighths

5. Each rectangle represents 1 sheet of paper. Estimate to draw lines to show how you would cut the paper into fractional units as indicated below.

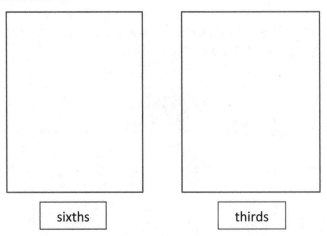

sixths   thirds

6. Yuri has a rope 12 meters long. He cuts it into pieces that are each 2 meters long. What fraction of the rope is one piece? Draw a picture. (You might fold a strip of paper to help you model the problem.)

7. Dawn bought 12 grams of chocolate. She ate half of the chocolate. How many grams of chocolate did she eat?

Lesson 1:   Specify and partition a whole into equal parts, identifying and
                              counting unit fractions using concrete models.

**EUREKA
MATH**™

Name _____   Date _____

1. Circle the strips that are folded to make equal parts.

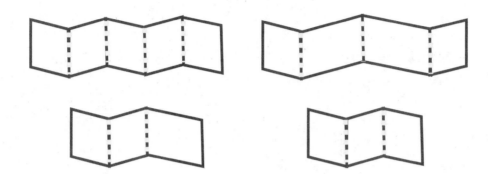

2.

   a. There are _____ equal parts in all. _____ are shaded.

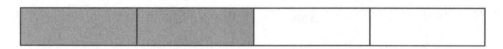

   b. There are _____ equal parts in all. _____ are shaded.

   c. There are _____ equal parts in all. _____ are shaded.

   d. There are _____ equal parts in all. _____ are shaded.

Lesson 2:   Specify and partition a whole into equal parts, identifying and
counting unit fractions by folding fraction strips.

©2015 Great Minds. eureka-math.org
G3-M5-SE-B3-1.3.1-01.2016

5

Use your fraction strips as tools to help you solve the following problems.

3. Noah, Pedro, and Sharon share a whole candy bar fairly. Which of your fraction strips shows how they each get an equal part? Draw the candy bar below. Then, label Sharon's fraction of the candy bar.

4. To make a garage for his toy truck, Zeno bends a rectangular piece of cardboard in half. He then bends each half in half again. Which of your fraction strips best matches this story?

   a. What fraction of the original cardboard is each part? Draw and label the matching fraction strip below.

   b. Zeno bends a different piece of cardboard in thirds. He then bends each third in half again. Which of your fraction strips best matches this story? Draw and label the matching fraction strip in the space below.

Lesson 2:   Specify and partition a whole into equal parts, identifying and counting unit fractions by folding fraction strips.

©2015 Great Minds. eureka-math.org
G3-M5-SE-B3-1.3.1-01.2016

**EUREKA
MATH™**

Name _____   Date _____

1. Circle the strips that are cut into equal parts.

2.

a. There are _____ equal parts in all. _____ is shaded.

b. There are _____ equal parts in all. _____ is shaded.

c. There are _____ equal parts in all. _____ is shaded.

d. There are _____ equal parts in all. _____ are shaded.

Lesson 2:   Specify and partition a whole into equal parts, identifying and counting unit fractions by folding fraction strips.

©2015 Great Minds. eureka-math.org
G3-M5-SE-B3-1.3.1-01.2016

7

3. Dylan plans to eat 1 fifth of his candy bar. His 4 friends want him to share the rest equally. Show how Dylan and his friends can each get an equal share of the candy bar.

4. Nasir baked a pie and cut it in fourths. He then cut each piece in half.

    a. What fraction of the original pie does each piece represent?

    b. Nasir ate 1 piece of pie on Tuesday and 2 pieces on Wednesday. What fraction of the original pie was not eaten?

Name _____   Date _____

1. Each shape is a whole divided into equal parts. Name the fractional unit, and then count and tell how many of those units are shaded. The first one is done for you.

_____Fourths_____

2 fourths are shaded.

_____   _____   _____

_____   _____   _____

2. Circle the shapes that are divided into equal parts. Write a sentence telling what *equal parts* means.

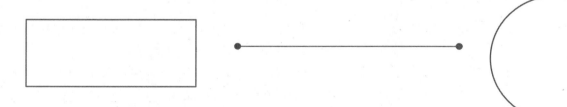

_____

_____

3. Each shape is 1 whole. Estimate to divide each into 4 equal parts. Name the fractional unit below.

Fractional unit:_____

Lesson 3:   Specify and partition a whole into equal parts, identifying and
counting unit fractions by drawing pictorial area models.

9

4. Each shape is 1 whole. Divide and shade to show the given fraction.

        1 half   1 sixth   1 third

5. Each shape is 1 whole. Estimate to divide each into equal parts (do not draw fourths). Divide each whole using a different fractional unit. Write the name of the fractional unit on the line below the shape.

_____   _____   _____

6. Charlotte wants to equally share a candy bar with 4 friends. Draw Charlotte's candy bar. Show how she can divide her candy bar so everyone gets an equal share. What fraction of the candy bar does each person receive?

                                                    Each person receives _____.

©2015 Great Minds. eureka-math.org
G3-M5-SE-B3-1.3.1-01.2016

Name _____   Date _____

1. Each shape is a whole divided into equal parts. Name the fractional unit, and then count and tell how many of those units are shaded. The first one is done for you.

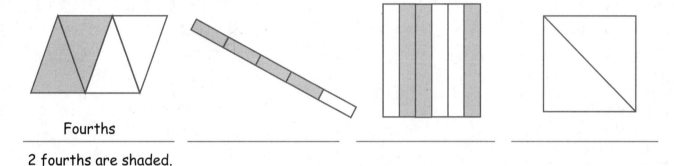

_____Fourths_____   _____   _____   _____

2 fourths are shaded.   _____   _____   _____

2. Each shape is 1 whole. Estimate to divide each into equal parts. Divide each whole using a different fractional unit. Write the name of the fractional unit on the line below the shape.

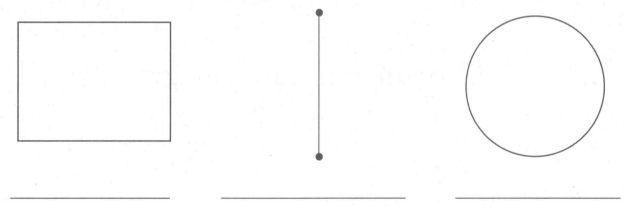

_____   _____   _____

3. Anita uses 1 sheet of paper to make a calendar showing each month of the year. Draw Anita's calendar. Show how she can divide her calendar so that each month is given the same space. What fraction of the calendar does each month receive?

Each month receives _____.

Lesson 3:   Specify and partition a whole into equal parts, identifying and counting unit fractions by drawing pictorial area models.

11

This page intentionally left blank

Name _____   Date _____

1. Draw a picture of the yellow strip at 3 (or 4) different stations. Shade and label 1 fractional unit of each.

2. Draw a picture of the brown bar at 3 (or 4) different stations. Shade and label 1 fractional unit of each.

3. Draw a picture of the square at 3 (or 4) different stations. Shade and label 1 fractional unit of each.

Lesson 4:   Represent and identify fractional parts of different wholes.

13

©2015 Great Minds. eureka-math.org
G3-M5-SE-B3-1.3.1-01.2016

4. Draw a picture of the clay at 3 (or 4) different stations. Shade and label 1 fractional unit of each.

5. Draw a picture of the water at 3 (or 4) different stations. Shade and label 1 fractional unit of each.

6. Extension: Draw a picture of the yarn at 3 (or 4) different stations.

Lesson 4:   Represent and identify fractional parts of different wholes.

Name _____   Date _____

Each shape is 1 whole. Estimate to equally partition the shape and shade to show the given fraction.

1. 1 half

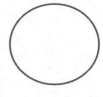      

A   B   C   D

2. 1 fourth

A   B   C   D

3. 1 third

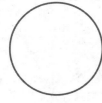   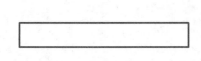      

A   B   C   D

4. Each of the shapes represents 1 whole. Match each shape to its fraction.

1 fifth

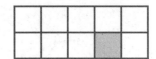

1 twelfth

1 third

1 fourth

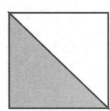

1 half

1 eighth

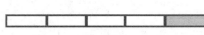

1 tenth

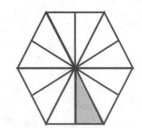

1 sixth

Lesson 4:   Represent and identify fractional parts of different wholes.

EUREKA MATH™

Name _____ Date _____

1. Fill in the chart. Each image is one whole.

	Total Number of Equal Parts	Total Number of Equal Parts Shaded	Unit Form	Fraction Form
a.				
b.				
c.				
d.				
e.				
f.				

EUREKA MATH

Lesson 5: Partition a whole into equal parts and define the equal parts to identify the unit fraction numerically.

17

©2015 Great Minds. eureka-math.org
G3-M5-SE-B3-1.3.1-01.2016

2. Andre's mom baked his 2 favorite cakes for his birthday party. The cakes were the exact same size. Andre cut his first cake into 8 pieces for him and his 7 friends. The picture below shows how he cut it. Did Andre cut the cake into eighths? Explain your answer.

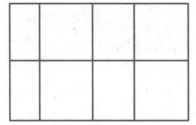

3. Two of Andre's friends came late to his party. They decide they will all share the second cake. Show how Andre can slice the second cake so that he and his nine friends can each get an equal amount with none leftover. What fraction of the second cake will they each receive?

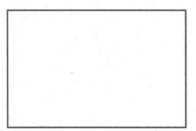

4. Andre thinks it's strange that $\frac{1}{10}$ of the cake would be less than $\frac{1}{8}$ of the cake since ten is bigger than eight. To explain to Andre, draw 2 identical rectangles to represent the cakes. Show 1 tenth shaded on one and 1 eighth shaded on the other. Label the unit fractions and explain to him which slice is bigger.

Lesson 5: Partition a whole into equal parts and define the equal parts to identify the unit fraction numerically.

EUREKA MATH™

Name _____   Date _____

1. Fill in the chart. Each image is one whole.

	Total Number of Equal Parts	Total Number of Equal Parts Shaded	Unit Form	Fraction Form
a.				
b.				
c.				
d.				
e.				

Lesson 5:   Partition a whole into equal parts and define the equal parts to identify the unit fraction numerically.

©2015 Great Minds. eureka-math.org
G3-M5-SE-B3-1.3.1-01.2016

2. This figure is divided into 6 parts. Are they sixths? Explain your answer.

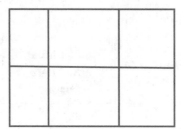

3. Terry and his 3 friends baked a pizza during his sleepover. They want to share the pizza equally. Show how Terry can slice the pizza so that he and his 3 friends can each get an equal amount with none left over.

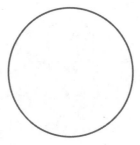

4. Draw two identical rectangles. Shade 1 seventh of one rectangle and 1 tenth of the other. Label the unit fractions. Use your rectangles to explain why $\frac{1}{7}$ is greater than $\frac{1}{10}$.

Lesson 5:   Partition a whole into equal parts and define the equal parts to identify the unit fraction numerically.

EUREKA MATH

Name _____   Date _____

1. Complete the number sentence. Estimate to partition each strip equally, write the unit fraction inside each unit, and shade the answer.

   Sample:

   2 thirds = $\dfrac{2}{3}$

   a. 3 fourths =

   b. 3 sevenths =

   c. 4 fifths =

   d. 2 sixths =

2. Mr. Stevens bought 8 liters of soda for a party. His guests drank 1 liter.

   a. What fraction of the soda did his guests drink?

   b. What fraction of the soda was left?

3. Fill in the chart.

	Total Number of Equal Parts	Total Number of Shaded Equal Parts	Unit Fraction	Fraction Shaded
Sample:	4	3	$\frac{1}{4}$	$\frac{3}{4}$
a.				
b.				
c.				
d.				
e.				

Lesson 6:   Build non-unit fractions less than one whole from unit fractions.

EUREKA
MATH

Name _____   Date _____

1. Complete the number sentence. Estimate to partition each strip equally, write the unit fraction inside each unit, and shade the answer.

   Sample:

   3 fourths = $\dfrac{3}{4}$

$\dfrac{1}{4}$	$\dfrac{1}{4}$	$\dfrac{1}{4}$	$\dfrac{1}{4}$

   a. 2 thirds =

   b. 5 sevenths =

   c. 3 fifths =

   d. 2 eighths =

2. Mr. Abney bought 6 kilograms of rice. He cooked 1 kilogram of it for dinner.

   a. What fraction of the rice did he cook for dinner?

   b. What fraction of the rice was left?

3. Fill in the chart.

	Total Number of Equal Parts	Total Number of Shaded Equal Parts	Unit Fraction	Fraction Shaded
Sample:	6	5	$\frac{1}{6}$	$\frac{5}{6}$
a.				
b.				
c.				
d.				

Lesson 6: Build non-unit fractions less than one whole from unit fractions.

EUREKA MATH

Name _____   Date _____

Whisper the fraction of the shape that is shaded. Then, match the shape to the amount that is <u>not</u> shaded.

1.

- 2 thirds

2.

- 6 sevenths

3.

- 4 fifths

4.

- 8 ninths

5.
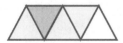

- 1 half

6.
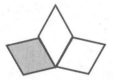

- 5 sixths

7.
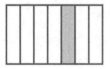

- 7 eighths

8.
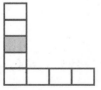

- 3 fourths

Lesson 7:   Identify and represent shaded and non-shaded parts of one whole as fractions.

25

9. a.   How many eighths are in 1 whole? _____

    b.   How many ninths are in 1 whole? _____

    c.   How many twelfths are in 1 whole? _____

10. Each strip represents 1 whole. Write a fraction to label the shaded and unshaded parts.

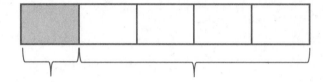

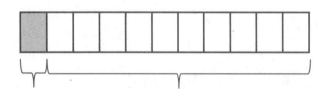

11. Avanti read 1 sixth of her book. What fraction of the book has she not read yet?

Lesson 7:   Identify and represent shaded and non-shaded parts of one whole as
                  fractions.

EUREKA
MATH™

Name _____   Date _____

Whisper the fraction of the shape that is shaded. Then, match the shape to the amount that is <u>not</u> shaded.

1.

    ▪ 9 tenths

2.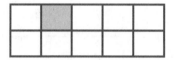

    ▪ 4 fifths

3.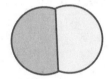

    ▪ 10 elevenths

4.

    ▪ 5 sixths

5.

    ▪ 1 half

6.

    ▪ 2 thirds

7.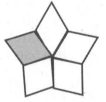

    ▪ 3 fourths

8.

    ▪ 6 sevenths

9. Each strip represents 1 whole. Write a fraction to label the shaded and unshaded parts.

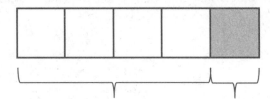

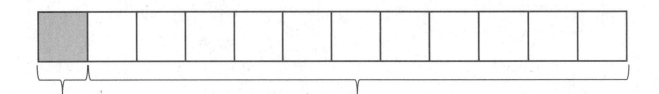

10. Carlia finished 1 fourth of her homework on Saturday. What fraction of her homework has she not finished? Draw and explain.

11. Jerome cooks 8 cups of oatmeal for his family. They eat 7 eighths of the oatmeal. What fraction of the oatmeal is uneaten? Draw and explain.

Lesson 7:   Identify and represent shaded and non-shaded parts of one whole as fractions.

EUREKA
MATH™

Name _____   Date _____

Show a number bond representing what is shaded and unshaded in each of the figures. Draw a different visual model that would be represented by the same number bond.

Sample:

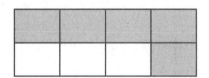 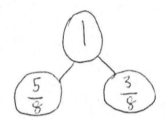 

1.

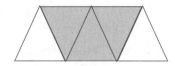

2.

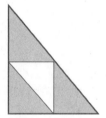

3.

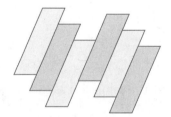

4.

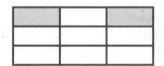

EUREKA
MATH™

5. Draw a number bond with 2 parts showing the shaded and unshaded fractions of each figure. Decompose both parts of the number bond into unit fractions.

a.

b.

c.

d.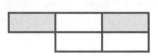

6. The chef put $\frac{1}{4}$ of the ground beef on the grill to make one hamburger and put the rest in the refrigerator. Draw a 2-part number bond showing the fraction of the ground beef on the grill and the fraction in the refrigerator. Draw a visual model of all the ground beef. Shade what is in the refrigerator.

a. What fraction of the ground beef was in the refrigerator?

b. How many more hamburgers can the chef make if he makes them all the same size as the first one?

c. Show the refrigerated ground beef broken into unit fractions on your number bond above.

EUREKA
MATH

Name _____ Date _____

Show a number bond representing what is shaded and unshaded in each of the figures. Draw a different visual model that would be represented by the same number bond.

Sample:

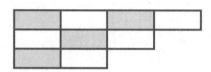

1.

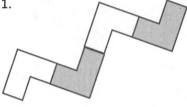

2.

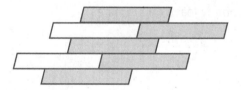

3.

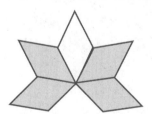

4.

Lesson 8:   Represent parts of one whole as fractions with number bonds.

31

EUREKA
MATH™

5. Draw a number bond with 2 parts showing the shaded and unshaded fractions of each figure. Decompose both parts of the number bond into unit fractions.

a.

b.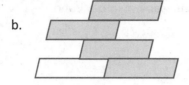

c.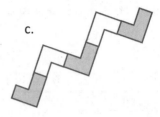

6. Johnny made a square peanut butter and jelly sandwich. He ate $\frac{1}{3}$ of it and left the rest on his plate. Draw a picture of Johnny's sandwich. Shade the part he left on his plate, and then draw a number bond that matches what you drew. What fraction of his sandwich did Johnny leave on his plate?

EUREKA MATH

Name _____   Date _____

1. Each figure represents 1 whole. Fill in the chart.

	Unit Fraction	Total Number of Units Shaded	Fraction Shaded
a. Sample:	$\frac{1}{2}$	5	$\frac{5}{2}$
b.			
c.			
d.			
e.			
f.			

Lesson 9:   Build and write fractions greater than one whole using unit fractions.

33

EUREKA MATH

©2015 Great Minds. eureka-math.org
G3-M5-SE-B3-1.3.1-01.2016

2. Estimate to draw and shade units on the fraction strips. Solve.

   Sample:

   5 thirds = $\frac{5}{3}$

   a. 8 sixths =

   ┌──────────────────────┐ ┌──────────────────────┐
   │   │ │   │
   └──────────────────────┘ └──────────────────────┘

   b. 7 fourths =

   ┌──────────────────────┐ ┌──────────────────────┐
   │   │ │   │
   └──────────────────────┘ └──────────────────────┘

   c. _____ = $\frac{6}{5}$

   ┌──────────────────────┐ ┌──────────────────────┐
   │   │ │   │
   └──────────────────────┘ └──────────────────────┘

   d. _____ = $\frac{5}{2}$

3. Mrs. Jawlik baked 2 pans of brownies. Draw the pans and estimate to partition each pan into 8 equal pieces.

   a. Mrs. Jawlik's children gobbled up 10 pieces. Shade the amount that was eaten.

   b. Write a fraction to show how many pans of brownies her children ate.

Lesson 9:   Build and write fractions greater than one whole using unit fractions.

EUREKA
MATH™

Name _____ Date _____

1. Each shape represents 1 whole. Fill in the chart.

	Unit Fraction	Total Number of Units Shaded	Fraction Shaded
a. Sample:	$\frac{1}{2}$	3	$\frac{3}{2}$
b.			
c.			
d.			
e.			
f.			

2. Estimate to draw and shade units on the fraction strips. Solve.

Sample:

7 fourths = $\frac{7}{4}$

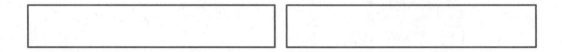

a. 5 thirds =

b. _____ = $\frac{9}{3}$

3. Reggie bought 2 candy bars. Draw the candy bars and estimate to partition each bar into 4 equal pieces.

a. Reggie ate 5 pieces. Shade the amount he ate.

b. Write a fraction to show how many candy bars Reggie ate.

Lesson 9: Build and write fractions greater than one whole using unit fractions.

EUREKA MATH

Name _____   Date _____

1. Each fraction strip is 1 whole. All the fraction strips are equal in length. Color 1 fractional unit in each strip. Then, answer the questions below.

$\frac{1}{2}$

$\frac{1}{4}$

$\frac{1}{8}$

$\frac{1}{3}$

$\frac{1}{6}$

2. Circle *less than* or *greater than*. Whisper the complete sentence.

a. $\frac{1}{2}$ is   less than $\frac{1}{4}$ b. $\frac{1}{6}$ is   less than $\frac{1}{2}$
       greater than   greater than

c. $\frac{1}{3}$ is   less than $\frac{1}{2}$ d. $\frac{1}{3}$ is   less than $\frac{1}{6}$
       greater than   greater than

e. $\frac{1}{8}$ is   less than $\frac{1}{6}$ f. $\frac{1}{8}$ is   less than $\frac{1}{4}$
       greater than   greater than

g. $\frac{1}{2}$ is   less than $\frac{1}{8}$ h. 9 eighths is   less than   2 halves
       greater than   greater than

3. Lily needs $\frac{1}{3}$ cup of oil and $\frac{1}{4}$ cup of water to make muffins. Will Lily use more oil or more water? Explain your answer using pictures, numbers, and words.

4. Use >, <, or = to compare.

   a. 1 third ◯ 1 fifth

   b. 1 seventh ◯ 1 fourth

   c. 1 sixth ◯ $\frac{1}{6}$

   d. 1 tenth ◯ $\frac{1}{12}$

   e. $\frac{1}{16}$ ◯ 1 eleventh

   f. 1 whole ◯ 2 halves

   Extension:

   g. $\frac{1}{8}$ ◯ 1 eighth ◯ $\frac{1}{6}$ ◯ $\frac{1}{3}$ ◯ 2 halves ◯ 1 whole

5. Your friend Eric says that $\frac{1}{6}$ is greater than $\frac{1}{5}$ because 6 is greater than 5. Is Eric correct? Use words and pictures to explain what happens to the size of a unit fraction when the number of parts gets larger.

Lesson 10: Compare unit fractions by reasoning about their size using fraction strips.

EUREKA MATH

Name _____ Date _____

1. Each fraction strip is 1 whole. All the fraction strips are equal in length. Color 1 fractional unit in each strip. Then, answer the questions below.

$\frac{1}{2}$

$\frac{1}{3}$

$\frac{1}{5}$

$\frac{1}{4}$

$\frac{1}{9}$

2. Circle *less than* or *greater than*. Whisper the complete sentence.

a. $\frac{1}{2}$ is   less than / greater than $\frac{1}{3}$

b. $\frac{1}{9}$ is   less than / greater than $\frac{1}{2}$

c. $\frac{1}{4}$ is   less than / greater than $\frac{1}{2}$

d. $\frac{1}{4}$ is   less than / greater than $\frac{1}{9}$

e. $\frac{1}{5}$ is   less than / greater than $\frac{1}{3}$

f. $\frac{1}{5}$ is   less than / greater than $\frac{1}{4}$

g. $\frac{1}{2}$ is   less than / greater than $\frac{1}{5}$

h. 6 fifths is   less than / greater than   3 thirds

EUREKA MATH

Lesson 10: Compare unit fractions by reasoning about their size using fraction strips.

39

©2015 Great Minds. eureka-math.org
G3-M5-SE-B3-1.3.1-01.2016

3. After his football game, Malik drinks $\frac{1}{2}$ liter of water and $\frac{1}{3}$ liter of juice. Did Malik drink more water or juice? Draw and estimate to partition. Explain your answer.

4. Use >, <, or = to compare.

    a. 1 fourth $\bigcirc$ 1 eighth

    b. 1 seventh $\bigcirc$ 1 fifth

    c. 1 eighth $\bigcirc$ $\frac{1}{8}$

    d. 1 twelfth $\bigcirc$ $\frac{1}{10}$

    e. $\frac{1}{15}$ $\bigcirc$ 1 thirteenth

    f. 3 thirds $\bigcirc$ 1 whole

5. Write a word problem about comparing fractions for your friends to solve. Be sure to show the solution so that your friends can check their work.

Lesson 10:   Compare unit fractions by reasoning about their size using fraction strips.

EUREKA
MATH™

Name _____   Date _____

Label the unit fraction. In each blank, draw and label the same whole with a shaded unit fraction that makes the sentence true. There is more than 1 correct way to make the sentence true.

Sample: $\dfrac{1}{4}$	is less than	$\dfrac{1}{2}$
1.	is greater than	
2.	is less than	
3.	is greater than	
4.	is less than	

EUREKA MATH

©2015 Great Minds. eureka-math.org
G3-M5-SE-B3-1.3.1-01.2016

5.	is greater than	
6.	is less than	
7.	is greater than	

8. Fill in the blank with a fraction to make the statement true, and draw a matching model.

$\frac{1}{4}$	is less than	$\frac{1}{2}$	is greater than

Lesson 11:   Compare unit fractions with different-sized models representing the whole.

EUREKA
MATH

9. Robert ate $\frac{1}{2}$ of a small pizza. Elizabeth ate $\frac{1}{4}$ of a large pizza. Elizabeth says, "My piece was larger than yours, so that means $\frac{1}{4} > \frac{1}{2}$." Is Elizabeth correct? Explain your answer.

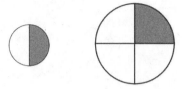

10. Manny and Daniel each ate $\frac{1}{2}$ of his candy, as shown below. Manny said he ate more candy than Daniel because his half is longer. Is he right? Explain your answer.

Manny's Candy Bar

Daniel's Candy Bar

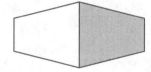

This page intentionally left blank

Name _____ Date _____

Label the unit fraction. In each blank, draw and label the same whole with a shaded unit fraction that makes the sentence true. There is more than 1 correct way to make the sentence true.

Sample: $\frac{1}{3}$	is less than	$\frac{1}{2}$
1.	is greater than	
2.	is less than	
3.	is greater than	
4.	is less than	

EUREKA MATH

Lesson 11:   Compare unit fractions with different-sized models representing the whole.

45

©2015 Great Minds. eureka-math.org
G3-M5-SE-B3-1.3.1-01.2016

5.	is greater than	
6.	is less than	
7.	is greater than	

8. Fill in the blank with a fraction to make the statement true. Draw a matching model.

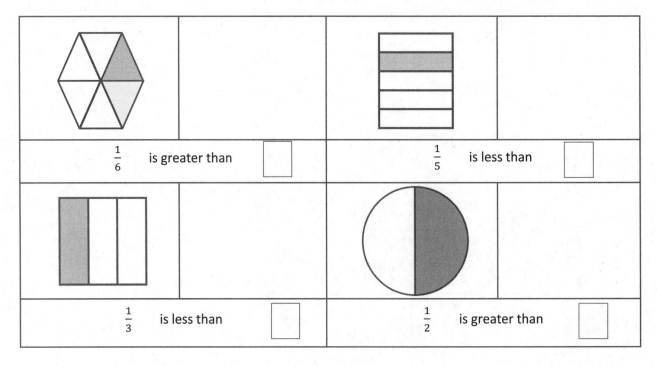

$\frac{1}{6}$ is greater than ☐		$\frac{1}{5}$ is less than ☐
$\frac{1}{3}$ is less than ☐		$\frac{1}{2}$ is greater than ☐

Lesson 11:   Compare unit fractions with different-sized models representing the whole.

EUREKA MATH

9. Debbie ate $\frac{1}{8}$ of a large brownie. Julian ate $\frac{1}{2}$ of a small brownie. Julian says, "I ate more than you because $\frac{1}{2} > \frac{1}{8}$."

a. Use pictures and words to explain Julian's mistake.

b. How could you change the problem so that Julian is correct? Use pictures and words to explain.

Lesson 11:   Compare unit fractions with different-sized models representing the whole.

47

©2015 Great Minds. eureka-math.org
G3-M5-SE-B3-1.3.1-01.2016

This page intentionally left blank

Name _____   Date _____

For each of the following:

- Draw a picture of the designated unit fraction copied to make at least two different wholes.
- Label the unit fractions.
- Label the whole as 1.
- Draw at least one number bond that matches a drawing.

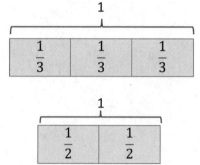

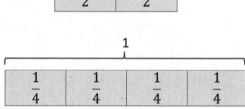

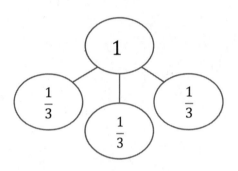

1. Yellow strip

2. Brown strip

EUREKA MATH™

Lesson 12:   Specify the corresponding whole when presented with one equal part.

49

3. Orange square

4. Yarn

5. Water

6. Clay

Lesson 12:   Specify the corresponding whole when presented with one equal part.

EUREKA
MATH™

Name _____   Date _____

Each shape represents the given unit fraction. Estimate to draw a possible whole.

1. $\frac{1}{2}$

2. $\frac{1}{6}$

3.   1 third

4.   1 fourth

Each shape represents the given unit fraction. Estimate to draw a possible whole, label the unit fractions, and draw a number bond that matches the drawing. The first one is done for you.

5. $\frac{1}{3}$

  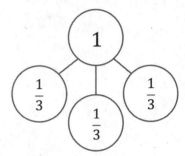

6. $\frac{1}{2}$

7. $\frac{1}{5}$

8. $\frac{1}{7}$

     Lesson 12:   Specify the corresponding whole when presented with one equal part.

EUREKA
MATH™

9. Evan and Yong used this shape ⬭, representing the unit fraction $\frac{1}{3}$, to draw 1 whole. Shania thinks both of them did it correctly. Do you agree with her? Explain your answer.

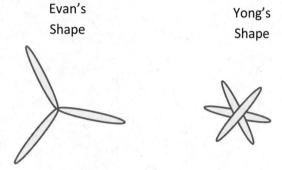

Evan's Shape   Yong's Shape

Lesson 12:   Specify the corresponding whole when presented with one equal part.

53

EUREKA
MATH

©2015 Great Minds. eureka-math.org
G3-M5-SE-B3-1.3.1-01.2016

This page intentionally left blank

Name _____   Date _____

The shape represents 1 whole. Write a unit fraction to describe the shaded part.	The shaded part represents 1 whole. Divide 1 whole to show the same unit fraction you wrote in Part (a).
1. a.	b.
2. a.	b.
3. a.	b.
4. a.	b.
5. a.	b.

Lesson 13:   Identify a shaded fractional part in different ways depending on the designation of the whole.

55

©2015 Great Minds. eureka-math.org
G3-M5-SE-B3-1.3.1-01.2016

6. Use the diagram below to complete the following statements.

Rope A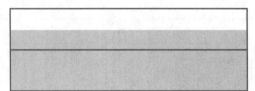

Rope B

Rope C

a. Rope _____ is $\frac{1}{2}$ the length of Rope B.

b. Rope _____ is $\frac{1}{2}$ the length of Rope A.

c. Rope C is $\frac{1}{4}$ the length of Rope _____.

d. If Rope B measures 1 m long, then Rope A is _____ m long, and Rope C is _____ m long.

e. If Rope A measures 1 m long, Rope B is _____ m long, and Rope C is _____ m long.

7. Ms. Fan drew the figure below on the board. She asked the class to name the shaded fraction. Charlie answered $\frac{3}{4}$. Janice answered $\frac{3}{2}$. Jenna thinks they're both right. With whom do you agree? Explain your thinking.

Lesson 13:   Identify a shaded fractional part in different ways depending on the designation of the whole.

EUREKA MATH™

Name _____ Date _____

The shape represents 1 whole. Write a fraction to describe the shaded part.	The shaded part represents 1 whole. Divide 1 whole to show the same unit fraction you wrote in Part (a).
1. a.	b.
2. a.	b.
3. a.	b.
4. a.	b.

EUREKA MATH

Lesson 13: Identify a shaded fractional part in different ways depending on the designation of the whole.

57

©2015 Great Minds. eureka-math.org
G3-M5-SE-B3-1.3.1-01.2016

5. Use the pictures below to complete the following statements.

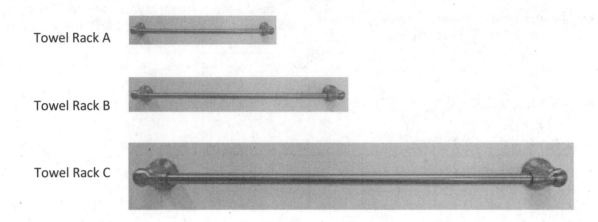

Towel Rack A

Towel Rack B

Towel Rack C

a. Towel Rack _____ is about $\frac{1}{2}$ the length of Towel Rack C.

b. Towel Rack _____ is about $\frac{1}{3}$ the length of Towel Rack C.

c. If Towel Rack C measures 6 ft long, then Towel Rack B is about _____ ft long, and Towel Rack A is about _____ ft long.

d. About how many copies of Towel Rack A equal the length of Towel Rack C? Draw number bonds to help you.

e. About how many copies of Towel Rack B equal the length of Towel Rack C? Draw number bonds to help you.

Identify a shaded fractional part in different ways depending on the designation of the whole.

EUREKA MATH

6. Draw 3 strings—B, C, and D—by following the directions below. String A is already drawn for you.

  ■ String B is $\frac{1}{3}$ of String A.

  ■ String C is $\frac{1}{2}$ of String B.

  ■ String D is $\frac{1}{3}$ of String C.

Extension: String E is 5 times the length of String D.

String A

Lesson 13:   Identify a shaded fractional part in different ways depending on the
designation of the whole.

©2015 Great Minds. eureka-math.org
G3-M5-SE-B3-1.3.1-01.2016

59

This page intentionally left blank

Name _____   Date _____

1. Draw a number bond for each fractional unit. Partition the fraction strip to show the unit fractions of the number bond. Use the fraction strip to help you label the fractions on the number line. Be sure to label the fractions at 0 and 1.

   a. Halves

   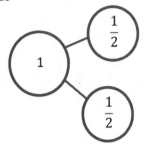

   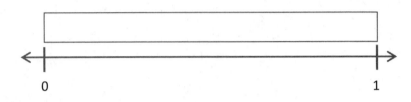

   b. Thirds

   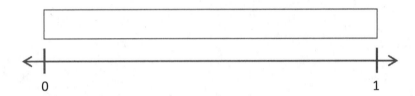

   c. Fourths

   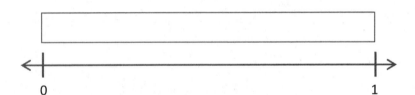

   d. Fifths

   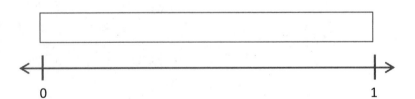

EUREKA
MATH™

Lesson 14:   Place fractions on a number line with endpoints 0 and 1.

61

2. Trevor needs to let his puppy outside every quarter (1 fourth) hour to potty train him. Draw and label a number line from 0 hours to 1 hour to show every 1 fourth hour. Include 0 fourths and 4 fourths hour. Label 0 hours and 1 hour, too.

3. A ribbon is 1 meter long. Mrs. Lee wants to sew a bead every $\frac{1}{5}$ meter. The first bead is at $\frac{1}{5}$ meter. The last bead is at 1 meter. Draw and label a number line from 0 meters to 1 meter to show where Mrs. Lee will sew beads. Label all the fractions, including 0 fifths and 5 fifths. Label 0 meters and 1 meter, too.

Lesson 14:   Place fractions on a number line with endpoints 0 and 1.

EUREKA MATH

Name _____ Date _____

1. Draw a number bond for each fractional unit. Partition the fraction strip to show the unit fractions of the number bond. Use the fraction strip to help you label the fractions on the number line. Be sure to label the fractions at 0 and 1.

a. Halves

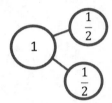

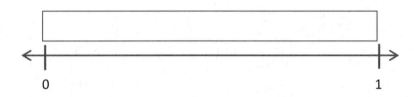

0   1

b. Eighths

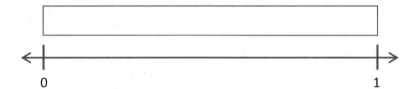

0   1

c. Fifths

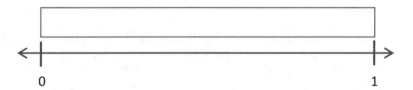

0   1

EUREKA
MATH™

Lesson 14:   Place fractions on a number line with endpoints 0 and 1.

63

2. Carter needs to wrap 7 presents. He lays the ribbon out flat and says, "If I make 6 equally spaced cuts, I'll have just enough pieces. I can use 1 piece for each package, and I won't have any pieces left over." Does he have enough pieces to wrap all the presents?

3. Mrs. Rivera is planting flowers in her 1-meter long rectangular plant box. She divides the plant box into sections $\frac{1}{9}$ meter in length, and plants 1 seed in each section. Draw and label a fraction strip representing the plant box from 0 meters to 1 meter. Represent each section where Mrs. Rivera will plant a seed. Label all the fractions.

    a. How many seeds will she be able to plant in 1 plant box?

    b. How many seeds will she be able to plant in 4 plant boxes?

    c. Draw a number line below your fraction strip and mark all the fractions.

Lesson 14:   Place fractions on a number line with endpoints 0 and 1.

                                                                    EUREKA
                                                                    MATH

Name _____ Date _____

1. Estimate to label the given fractions on the number line. Be sure to label the fractions at 0 and 1. Write the fractions above the number line. Draw a number bond to match your number line.

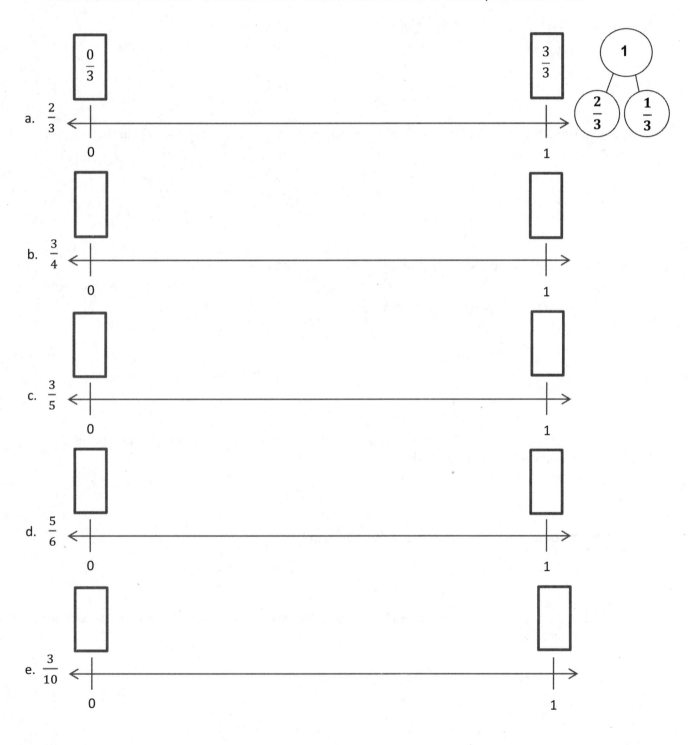

EUREKA
MATH™

Lesson 15:   Place any fraction on a number line with endpoints 0 and 1.

65

©2015 Great Minds. eureka-math.org
G3-M5-SE-B3-1.3.1-01.2016

2. Draw a number line. Use a fraction strip to locate 0 and 1. Fold the strip to make 8 equal parts. Use the strip to measure and label your number line with eighths.

Count up from 0 eighths to 8 eighths on your number line. Touch each number with your finger as you count.

3. For his boat, James stretched out a rope with 5 equally spaced knots as shown.

a. Starting at the first knot and ending at the last knot, how many equal parts are formed by the 5 knots? Label each fraction at the knot.

b. What fraction of the rope is labeled at the third knot?

c. What if the rope had 6 equally spaced knots along the same length? What fraction of the rope would be measured by the first 2 knots?

Lesson 15:   Place any fraction on a number line with endpoints 0 and 1.

EUREKA
MATH™

Name _____ Date _____

1. Estimate to label the given fractions on the number line. Be sure to label the fractions at 0 and 1. Write the fractions above the number line. Draw a number bond to match your number line. The first one is done for you.

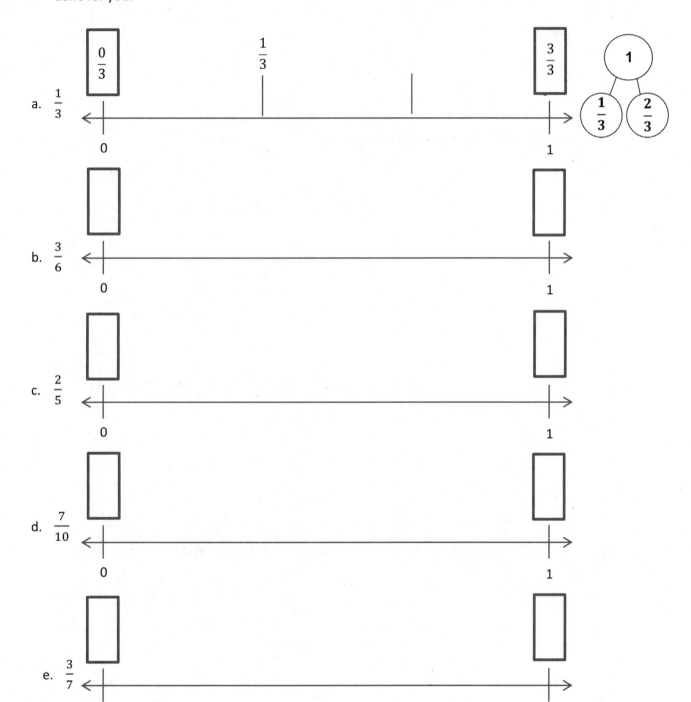

Lesson 15: Place any fraction on a number line with endpoints 0 and 1.

67

EUREKA
MATH™

©2015 Great Minds. eureka-math.org
G3-M5-SE-B3-1.3.1-01.2016

2. Henry has 5 dimes. Ben has 9 dimes. Tina has 2 dimes.

    a.   Write the value of each person's money as a fraction of a dollar:

         Henry:

         Ben:

         Tina:

    b.   Estimate to place each fraction on the number line.

         $0 $1

3. Draw a number line. Use a fraction strip to locate 0 and 1. Fold the strip to make 8 equal parts.

    a.   Use the strip to measure and label your number line with eighths.

    b.   Count up from 0 eighths to 8 eighths on your number line. Touch each number with your finger as you count.

EUREKA
MATH™

Name _____   Date _____

1. Estimate to equally partition and label the fractions on the number line. Label the wholes as fractions, and box them. The first one is done for you.

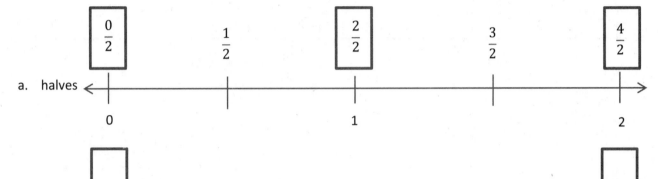

a. halves

b. thirds

c. halves

d. fourths

e. thirds

EUREKA MATH

Lesson 16:   Place whole number fractions and fractions between whole numbers on the number line.

69

2. Partition each whole into fifths. Label each fraction. Count up as you go. Box the fractions that are located at the same points as whole numbers.

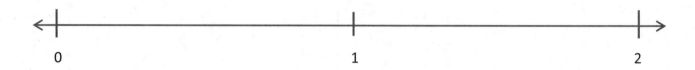

3. Partition each whole into thirds. Label each fraction. Count up as you go. Box the fractions that are located at the same points as whole numbers.

4. Draw a number line with endpoints 0 and 3. Label the wholes. Partition each whole into fourths. Label all the fractions from 0 to 3. Box the fractions that are located at the same points as whole numbers. Use a separate paper if you need more space.

Lesson 16:   Place whole number fractions and fractions between whole numbers on the number line.

EUREKA
MATH™

Name _____   Date _____

1. Estimate to equally partition and label the fractions on the number line. Label the wholes as fractions, and box them. The first one is done for you.

a. thirds

b. eighths

c. fourths

d. halves

e. fifths

EUREKA MATH

Lesson 16:   Place whole number fractions and fractions between whole numbers   71
on the number line.

©2015 Great Minds. eureka-math.org
G3-M5-SE-B3-1.3.1-01.2016

2. Partition each whole into sixths. Label each fraction. Count up as you go. Box the fractions that are located at the same points as whole numbers.

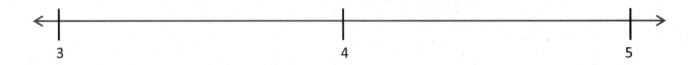

3. Partition each whole into halves. Label each fraction. Count up as you go. Box the fractions that are located at the same points as whole numbers.

4. Draw a number line with endpoints 0 and 3. Label the wholes. Partition each whole into fifths. Label all the fractions from 0 to 3. Box the fractions that are located at the same points as whole numbers. Use a separate paper if you need more space.

Lesson 16:   Place whole number fractions and fractions between whole numbers on the number line.

©2015 Great Minds. eureka-math.org
G3-M5-SE-B3-1.3.1-01.2016

EUREKA MATH™

Name _____   Date _____

1. Locate and label the following fractions on the number line.

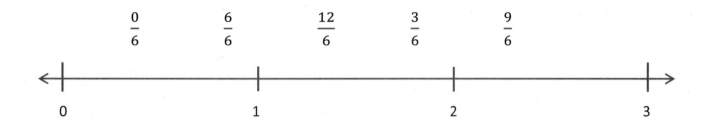

2. Locate and label the following fractions on the number line.

3. Locate and label the following fractions on the number line.

$$\frac{18}{3} \qquad \frac{14}{3} \qquad \frac{9}{3} \qquad \frac{11}{3} \qquad \frac{6}{3}$$

4. For a measurement project in math class, students measured the lengths of their pinky fingers. Alex's measured 2 inches long. Jerimiah's pinky finger was $\frac{7}{4}$ inches long. Whose finger is longer? Draw a number line to help prove your answer.

5. Marcy ran 4 kilometers after school. She stopped to tie her shoelace at $\frac{7}{5}$ kilometers. Then, she stopped to switch songs on her iPod at $\frac{12}{5}$ kilometers. Draw a number line showing Marcy's run. Include her starting and finishing points and the 2 places where she stopped.

 Lesson 17: Practice placing various fractions on the number line.

EUREKA MATH

Name _____   Date _____

1. Locate and label the following fractions on the number line.

$$\frac{1}{2} \qquad \frac{4}{2} \qquad \frac{5}{2}$$

0   1   2   3

2. Locate and label the following fractions on the number line.

$$\frac{11}{3} \qquad \frac{6}{3} \qquad \frac{8}{3}$$

2   3   4

3. Locate and label the following fractions on the number line.

$$\frac{20}{4} \qquad \frac{13}{4} \qquad \frac{23}{4}$$

3   4   5   6

4. Wayne went on a 4-kilometer hike. He took a break at $\frac{4}{3}$ kilometers. He took a drink of water at $\frac{10}{3}$ kilometers. Show Wayne's hike on the number line. Include his starting and finishing place and the 2 points where he stopped.

5. Ali wants to buy a piano. The piano measures $\frac{19}{4}$ feet long. She has a space 5 feet long for the piano in her house. Does she have enough room? Draw a number line to show, and explain your answer.

4 ft   5 ft

EUREKA
MATH™

Name _____   Date _____

Place the two fractions on the number line. Circle the fraction with the distance closest to 0. Then, compare using >, <, or =. The first problem is done for you.

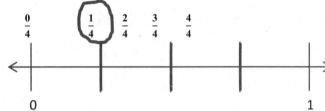

1.

2.   

3.   

4.   

5.   

EUREKA MATH

Lesson 18:   Compare fractions and whole numbers on the number line by reasoning about their distance from 0.

77

©2015 Great Minds. eureka-math.org
G3-M5-SE-B3-1.3.1-01.2016

6. JoAnn and Lupe live straight down the street from their school. JoAnn walks $\frac{5}{6}$ miles and Lupe walks $\frac{7}{8}$ miles home from school every day. Draw a number line to model how far each girl walks. Who walks the least? Explain how you know using pictures, numbers, and words.

7. Cheryl cuts 2 pieces of thread. The blue thread is $\frac{5}{4}$ meters long. The red thread is $\frac{4}{5}$ meters long. Draw a number line to model the length of each piece of thread. Which piece of thread is shorter? Explain how you know using pictures, numbers, and words.

8. Brandon makes homemade spaghetti. He measures 3 noodles. One measures $\frac{7}{8}$ feet, the second is $\frac{7}{4}$ feet, and the third is $\frac{4}{2}$ feet long. Draw a number line to model the length of each piece of spaghetti. Write a number sentence using <, >, or = to compare the pieces. Explain using pictures, numbers, and words.

Compare fractions and whole numbers on the number line by
reasoning about their distance from 0.

EUREKA
MATH™

Name _____   Date _____

Place the two fractions on the number line. Circle the fraction with the distance closest to 0. Then, compare using >, <, or =.

1. $\dfrac{1}{3}$ ◯ $\dfrac{2}{3}$

2. $\dfrac{4}{6}$ ◯ $\dfrac{1}{6}$

3. $\dfrac{1}{4}$ ◯ $\dfrac{1}{8}$

4. $\dfrac{4}{5}$ ◯ $\dfrac{4}{10}$

5. $\dfrac{8}{6}$ ◯ $\dfrac{5}{3}$

EUREKA MATH

Lesson 18:   Compare fractions and whole numbers on the number line by reasoning about their distance from 0.

79

©2015 Great Minds. eureka-math.org
G3-M5-SE-B3-1.3.1-01.2016

6. Liz and Jay each have a piece of string. Liz's string is $\frac{4}{6}$ yards long, and Jay's string is $\frac{5}{7}$ yards long. Whose string is longer? Draw a number line to model the length of both strings. Explain the comparison using pictures, numbers, and words.

7. In a long jump competition, Wendy jumped $\frac{9}{10}$ meters, and Judy jumped $\frac{10}{9}$ meters. Draw a number line to model the distance of each girl's long jump. Who jumped the shorter distance? Explain how you know using pictures, numbers, and words.

8. Nikki has 3 pieces of yarn. The first piece is $\frac{5}{6}$ feet long, the second piece is $\frac{5}{3}$ feet long, and the third piece is $\frac{3}{2}$ feet long. She wants to arrange them from the shortest to the longest. Draw a number line to model the length of each piece of yarn. Write a number sentence using <, >, or = to compare the pieces. Explain using pictures, numbers, and words.

Lesson 18:   Compare fractions and whole numbers on the number line by
             reasoning about their distance from 0.

EUREKA
MATH

Name _____   Date _____

1. Divide each number line into the given fractional unit. Then, place the fractions. Write each whole as a fraction.

   a. halves $\frac{3}{2}$ $\frac{5}{2}$ $\frac{4}{2}$

   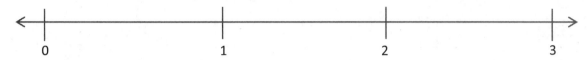

   b. fourths $\frac{9}{4}$ $\frac{11}{4}$ $\frac{6}{4}$

   c. eighths $\frac{24}{8}$ $\frac{19}{8}$ $\frac{16}{8}$

2. Use the number lines above to compare the following fractions using >, <, or =.

   $\frac{6}{4}$ ◯ $\frac{9}{4}$ $\frac{3}{2}$ ◯ $\frac{5}{2}$ $\frac{19}{8}$ ◯ $\frac{16}{8}$

   $\frac{16}{8}$ ◯ $\frac{3}{2}$ $\frac{9}{4}$ ◯ $\frac{19}{8}$ $\frac{4}{2}$ ◯ $\frac{16}{8}$

   $\frac{6}{4}$ ◯ $\frac{16}{8}$ $\frac{5}{2}$ ◯ $\frac{9}{4}$ $\frac{24}{8}$ ◯ $\frac{11}{4}$

EUREKA MATH

Lesson 19:   Understand distance and position on the number line as strategies for comparing fractions. (Optional)

81

3. Choose a *greater than* comparison you made in Problem 2. Use pictures, numbers, and words to explain how you made that comparison.

4. Choose a *less than* comparison you made in Problem 2. Use pictures, numbers, and words to explain a different way of thinking about the comparison than what you wrote in Problem 3.

5. Choose an *equal to* comparison you made in Problem 2. Use pictures, numbers, and words to explain two ways that you can prove your comparison is true.

Lesson 19:   Understand distance and position on the number line as strategies for comparing fractions. (Optional)

EUREKA
MATH™

Name _____   Date _____

1. Divide each number line into the given fractional unit. Then, place the fractions. Write each whole as a fraction.

   a. thirds $\frac{6}{3}$ $\frac{5}{3}$ $\frac{8}{3}$

   b. sixths $\frac{10}{6}$ $\frac{18}{6}$ $\frac{15}{6}$

   c. fifths $\frac{14}{5}$ $\frac{7}{5}$ $\frac{11}{5}$

2. Use the number lines above to compare the following fractions using >, <, or =.

   $\frac{17}{6}$ ◯ $\frac{15}{6}$ $\frac{7}{3}$ ◯ $\frac{9}{3}$ $\frac{11}{5}$ ◯ $\frac{8}{5}$

   $\frac{4}{3}$ ◯ $\frac{8}{6}$ $\frac{13}{6}$ ◯ $\frac{8}{3}$ $\frac{11}{6}$ ◯ $\frac{5}{3}$

   $\frac{10}{6}$ ◯ $\frac{3}{3}$ $\frac{6}{3}$ ◯ $\frac{12}{6}$ $\frac{15}{5}$ ◯ $\frac{5}{3}$

EUREKA
MATH™

Lesson 19:   Understand distance and position on the number line as strategies for
              comparing fractions. (Optional)

©2015 Great Minds. eureka-math.org
G3-M5-SE-B3-1.3.1-01.2016

83

3. Use fractions from the number lines in Problem 1. Complete the sentence. Use words, pictures, or numbers to explain how you made that comparison.

_____ is *greater than* _____.

4. Use fractions from the number lines in Problem 1. Complete the sentence. Use words, pictures, or numbers to explain how you made that comparison.

_____ is *less than* _____.

5. Use fractions from the number lines in Problem 1. Complete the sentence. Use words, pictures, or numbers to explain how you made that comparison.

_____ is *equal to* _____.

Lesson 19:   Understand distance and position on the number line as strategies for comparing fractions. (Optional)

EUREKA
MATH™

Name _____   Date _____

1. Label what fraction of each shape is shaded. Then, circle the fractions that are equal.

    a.

    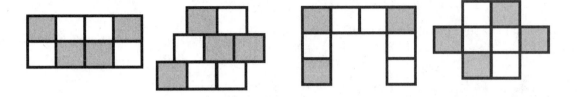

    b.

    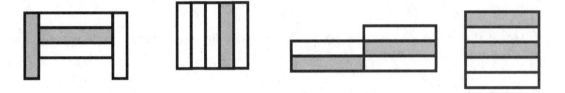

    c.

    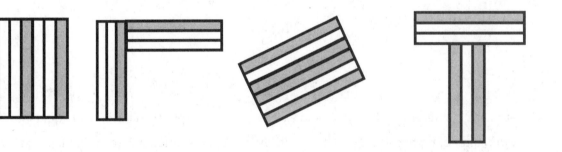

2. Label the shaded fraction. Draw 2 different representations of the same fractional amount.

    a.

    b.

    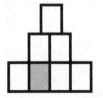

EUREKA
MATH™

Lesson 20:   Recognize and show that equivalent fractions have the same size,
              though not necessarily the same shape.

©2015 Great Minds. eureka-math.org
G3-M5-SE-B3-1.3.1-01.2016

85

3. Ann has 6 small square pieces of paper. 2 squares are grey. Ann cuts the 2 grey squares in half with a diagonal line from one corner to the other.

    a. What shapes does she have now?

    b. How many of each shape does she have?

    c. Use all the shapes with no overlaps. Draw at least 2 different ways Ann's set of shapes might look. What fraction of the figure is grey?

4. Laura has 2 different beakers that hold exactly 1 liter. She pours $\frac{1}{2}$ liter of blue liquid into Beaker A. She pours $\frac{1}{2}$ liter of orange liquid into Beaker B. Susan says the amounts are not equal. Cristina says they are. Explain who you think is correct and why.

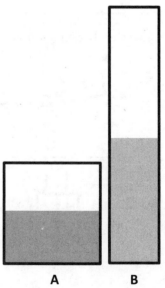

Recognize and show that equivalent fractions have the same size, though not necessarily the same shape.

EUREKA MATH

Name _____   Date _____

1. Label the shaded fraction. Draw 2 different representations of the same fractional amount.

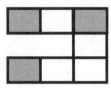

2. These two shapes both show $\frac{4}{5}$.

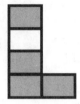 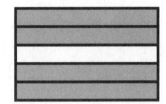

     a. Are the shapes equivalent? Why or why not?

     b. Draw two different representations of $\frac{4}{5}$ that are equivalent.

3. Diana ran a quarter mile straight down the street. Becky ran a quarter mile on a track. Who ran more? Explain your thinking.

     Diana

     Becky   ◯

Lesson 20:   Recognize and show that equivalent fractions have the same size, though not necessarily the same shape.

87

©2015 Great Minds. eureka-math.org
G3-M5-SE-B3-1.3.1-01.2016

This page intentionally left blank

Name _____   Date _____

1. Use the fractional units on the left to count up on the number line. Label the missing fractions on the blanks.

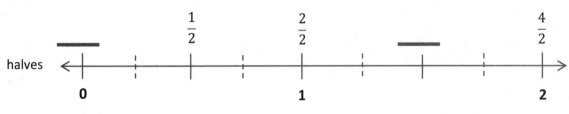

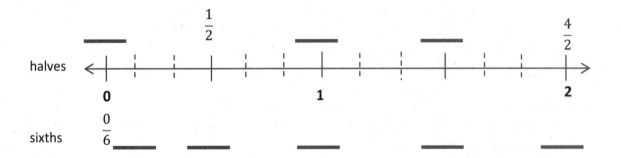

2. Use the number lines above to:

   ▪   Color fractions equal to 1 half blue.

   ▪   Color fractions equal to 1 yellow.

   ▪   Color fractions equal to 3 halves green.

   ▪   Color fractions equal to 2 red.

3. Use the number lines above to make the number sentences true.

$$\frac{2}{4} = \frac{}{6} \qquad \frac{6}{6} = \frac{2}{} = \frac{}{} \qquad \frac{3}{2} = \frac{}{6} = \frac{6}{}$$

EUREKA
MATH™

Lesson 21:   Recognize and show that equivalent fractions refer to the same point
on the number line.

89

©2015 Great Minds. eureka-math.org
G3-M5-SE-B3-1.3.1-01.2016

4. Jack and Jill use rain gauges the same size and shape to measure rain on the top of a hill. Jack uses a rain gauge marked in fourths of an inch. Jill's gauge measures rain in eighths of an inch. On Thursday, Jack's gauge measured $\frac{2}{4}$ inches of rain. They both had the same amount of water, so what was the reading on Jill's gauge Thursday? Draw a number line to help explain your thinking.

5. Jack and Jill's baby brother Rosco also had a gauge the same size and shape on the same hill. He told Jack and Jill that there had been $\frac{1}{2}$ inch of rain on Thursday. Is he right? Why or why not? Use words and a number line to explain your answer.

Lesson 21:   Recognize and show that equivalent fractions refer to the same point on the number line.

EUREKA
MATH™

Name _____ Date _____

1. Use the fractional units on the left to count up on the number line. Label the missing fractions on the blanks.

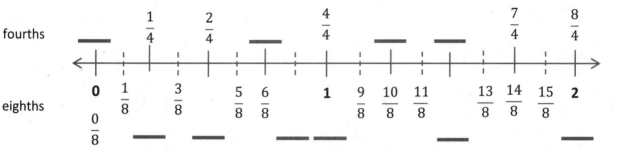

2. Use the number lines above to:

   ▪ Color fractions equal to 1 purple.
   ▪ Color fractions equal to 2 fourths yellow.
   ▪ Color fractions equal to 2 blue.
   ▪ Color fractions equal to 5 thirds green.
   ▪ Write a pair of fractions that are equivalent.

   _____ = _____

EUREKA
MATH™

Lesson 21: Recognize and show that equivalent fractions refer to the same point
on the number line.

91

©2015 Great Minds. eureka-math.org
G3-M5-SE-B3-1.3.1-01.2016

3. Use the number lines on the previous page to make the number sentences true.

$$\frac{1}{4} = \frac{}{8} \qquad \frac{6}{4} = \frac{12}{} \qquad \frac{2}{3} = \frac{}{6}$$

$$\frac{6}{3} = \frac{12}{} \qquad \frac{3}{3} = \frac{}{6} \qquad 2 = \frac{8}{4} = \frac{}{8}$$

4. Mr. Fairfax ordered 3 large pizzas for a class party. Group A ate $\frac{6}{6}$ of the first pizza, and Group B ate $\frac{8}{6}$ of the remaining pizza. During the party, the class discussed which group ate more pizza.

a. Did Group A or B eat more pizza? Use words and pictures to explain your answer to the class.

b. Later, Group C ate all remaining slices of pizza. What fraction of the pizza did group C eat? Use words and pictures to explain your answer.

Lesson 21:   Recognize and show that equivalent fractions refer to the same point on the number line.

©2015 Great Minds. eureka-math.org
G3-M5-SE-B3-1.3.1-01.2016

EUREKA
MATH™

Name _____ Date _____

1. Write the shaded fraction of each figure on the blank. Then, draw a line to match the equivalent fractions.

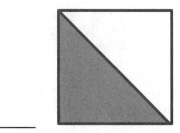
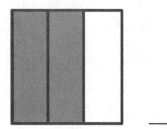

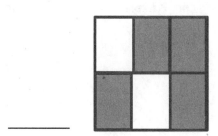
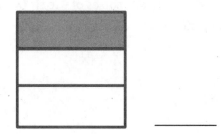

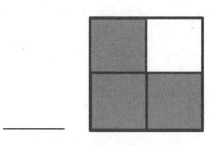
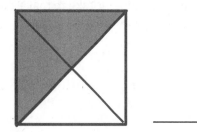

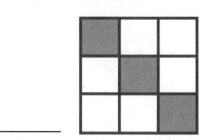
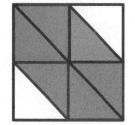

EUREKA MATH

Lesson 22: Generate simple equivalent fractions by using visual fraction models and the number line.

©2015 Great Minds. eureka-math.org
G3-M5-SE-B3-1.3.1-01.2016

93

2. Write the missing parts of the fractions.

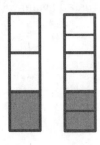   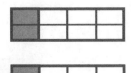   

$$\frac{1}{3} = \frac{}{6} \qquad \frac{2}{} = \frac{1}{4} \qquad \frac{4}{8} = \frac{8}{}$$

3. Why does it take 2 copies of $\frac{1}{8}$ to show the same amount as 1 copy of $\frac{1}{4}$? Explain your answer in words and pictures.

4. How many sixths does it take to make the same amount as $\frac{1}{3}$? Explain your answer in words and pictures.

5. Why does it take 10 copies of 1 sixth to make the same amount as 5 copies of 1 third? Explain your answer in words and pictures.

      Lesson 22:   Generate simple equivalent fractions by using visual fraction models and the number line.   **EUREKA MATH**

Name _____ Date _____

1. Write the shaded fraction of each figure on the blank. Then, draw a line to match the equivalent fractions.

EUREKA MATH

Lesson 22: Generate simple equivalent fractions by using visual fraction models and the number line.

95

©2015 Great Minds. eureka-math.org
G3-M5-SE-B3-1.3.1-01.2016

2. Complete the fractions to make true statements.

$$\frac{1}{2} = \frac{4}{}$$

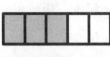
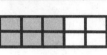

$$\frac{3}{5} = \frac{}{10}$$

$$\frac{3}{9} = \frac{6}{}$$

3. Why does it take 3 copies of $\frac{1}{6}$ to show the same amount as 1 copy of $\frac{1}{2}$? Explain your answer in words and pictures.

4. How many ninths does it take to make the same amount as $\frac{1}{3}$? Explain your answer in words and pictures.

5. A pie was cut into 8 equal slices. If Ruben ate $\frac{3}{4}$ of the pie, how many slices did he eat? Explain your answer using a number line and words.

Lesson 22:   Generate simple equivalent fractions by using visual fraction models
and the number line.

EUREKA
MATH™

Name _____   Date _____

1. On the number line above, use a red colored pencil to divide each whole into fourths, and label each fraction above the line. Use a fraction strip to help you estimate, if necessary.

2. On the number line above, use a blue colored pencil to divide each whole into eighths, and label each fraction below the line. Refold your fraction strip from Problem 1 to help you estimate.

3. List the fractions that name the same place on the number line.

4. Using your number line to help, what red fraction and what blue fraction would be equal to $\frac{7}{2}$? Draw the part of the number line below that would include these fractions, and label it.

EUREKA MATH   **Lesson 23:**   Generate simple equivalent fractions by using visual fraction models and the number line.   97

©2015 Great Minds. eureka-math.org
G3-M5-SE-B3-1.3.1-01.2016

5. Write two different fractions for the dot on the number line. You may use halves, thirds, fourths, fifths, sixths, or eighths. Use fraction strips to help you, if necessary.

_____ = _____

_____ = _____

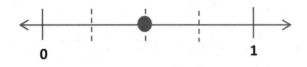

_____ = _____

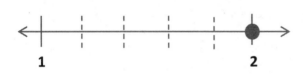

_____ = _____

6. Cameron and Terrance plan to run in the city race on Saturday. Cameron has decided that he will divide his race into 3 equal parts and will stop to rest after running 2 of them. Terrance divides his race into 6 equal parts and will stop and rest after running 2 of them. Will the boys rest at the same spot in the race? Why or why not? Draw a number line to explain your answer.

EUREKA
MATH™

Name _____ Date _____

0   1   2   3

1. On the number line above, use a colored pencil to divide each whole into thirds and label each fraction above the line.

2. On the number line above, use a different colored pencil to divide each whole into sixths and label each fraction below the line.

3. Write the fractions that name the same place on the number line.

4. Using your number line to help, name the fraction equivalent to $\frac{20}{6}$. Name the fraction equivalent to $\frac{12}{3}$. Draw the part of the number line that would include these fractions below, and label it.

$$\frac{20}{6} = \frac{}{3} \qquad\qquad \frac{12}{3} = \frac{}{6}$$

EUREKA MATH   **Lesson 23:**   Generate simple equivalent fractions by using visual fraction models and the number line.   99

©2015 Great Minds. eureka-math.org
G3-M5-SE-B3-1.3.1-01.2016

5. Write two different fraction names for the dot on the number line. You may use halves, thirds, fourths, fifths, sixths, eighths, or tenths.

_____ = _____

_____ = _____

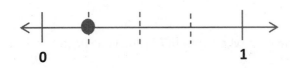

_____ = _____

_____ = _____

6. Danielle and Mandy each ordered a large pizza for dinner. Danielle's pizza was cut into sixths, and Mandy's pizza was cut into twelfths. Danielle ate 2 sixths of her pizza. If Mandy wants to eat the same amount of pizza as Danielle, how many slices of pizza will she have to eat? Write the answer as a fraction. Draw a number line to explain your answer.

     Lesson 23:   Generate simple equivalent fractions by using visual fraction models and the number line.

©2015 Great Minds. eureka-math.org
G3-M5-SE-B3-1.3.1-01.2016

EUREKA
MATH™

Name _____   Date _____

1. Complete the number bond as indicated by the fractional unit. Partition the number line into the given fractional unit, and label the fractions. Rename 0 and 1 as fractions of the given unit. The first one is done for you.

Halves

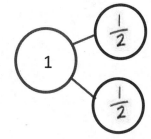

Thirds

Fourths

Fifths

2. Circle all the fractions in Problem 1 that are equal to 1. Write them in a number sentence below.

$\dfrac{2}{2}$ = _____ = _____ = _____

3. What pattern do you notice in the fractions that are equivalent to 1?

4. Taylor took his little brother to get pizza. Each boy ordered a small pizza. Taylor's pizza was cut in fourths, and his brother's was cut in thirds. After they had both eaten all of their pizza, Taylor's little brother said, "Hey that was no fair! You got more than me! You got 4 pieces, and I only got 3."

Should Taylor's little brother be mad? What could you say to explain the situation to him? Use words, pictures, or a number line.

      Lesson 24:   Express whole numbers as fractions and recognize equivalence with different units.

EUREKA
MATH™

Name _____   Date _____

1. Complete the number bond as indicated by the fractional unit. Partition the number line into the given fractional unit, and label the fractions. Rename 0 and 1 as fractions of the given unit.

Fifths

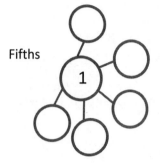

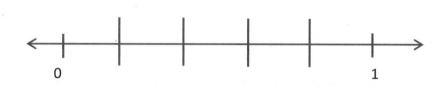

Sixths

Sevenths

Eighths

EUREKA MATH

Lesson 24:   Express whole numbers as fractions and recognize equivalence with different units.

103

2. Circle all the fractions in Problem 1 that are equal to 1. Write them in a number sentence below.

$\dfrac{5}{5}$ = _____ = _____ = _____

3. What pattern do you notice in the fractions that are equivalent to 1? Following this pattern, how would you represent ninths as 1 whole?

4. In Art class, Mr. Joselyn gave everyone a 1-foot stick to measure and cut. Vivian measured and cut her stick into 5 equal pieces. Scott measured and cut his into 7 equal pieces. Scott said to Vivian, "The total length of my stick is longer than yours because I have 7 pieces, and you only have 5." Is Scott correct? Use words, pictures, or a number line to help you explain.

EUREKA
MATH

Name _____ Date _____

1. Label the following models as a fraction inside the dotted box. The first one has been done for you.

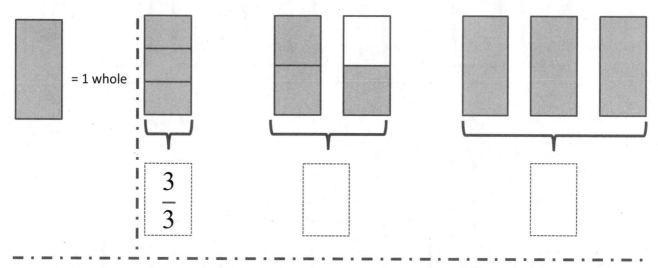

$$\frac{3}{3}$$

EUREKA MATH

Lesson 25: Express whole number fractions on the number line when the unit interval is 1.

105

©2015 Great Minds. eureka-math.org
G3-M5-SE-B3-1.3.1-01.2016

2. Fill in the missing whole numbers in the boxes below the number line. Rename the whole numbers as fractions in the boxes above the number line.

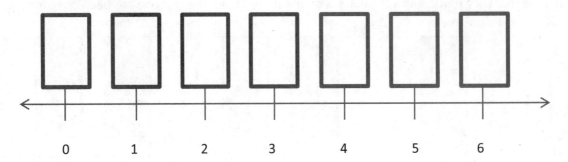

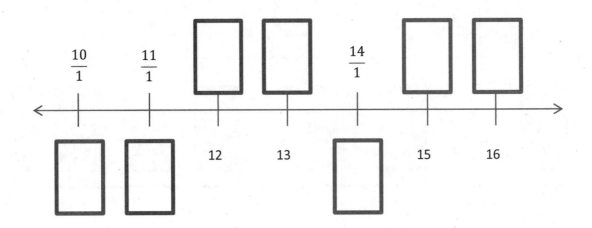

3. Explain the difference between these two fractions with words and pictures.

$$\frac{2}{1} \qquad\qquad \frac{2}{2}$$

Lesson 25:   Express whole number fractions on the number line when the unit interval is 1.

EUREKA MATH

Name _____ Date _____

1. Label the following models as fractions inside the boxes.

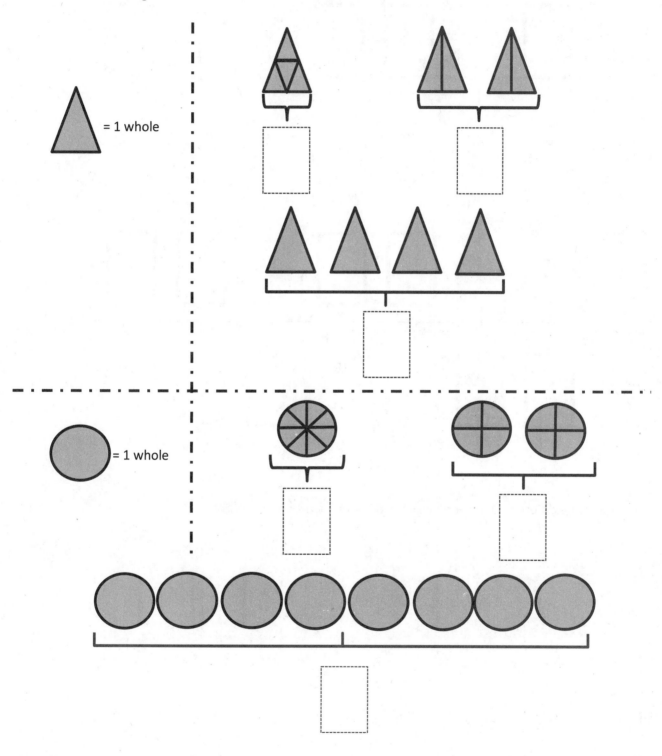

EUREKA MATH

Lesson 25: Express whole number fractions on the number line when the unit interval is 1.

107

2. Fill in the missing whole numbers in the boxes below the number line. Rename the wholes as fractions in the boxes above the number line.

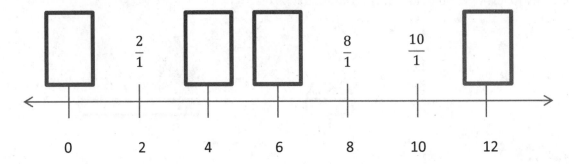

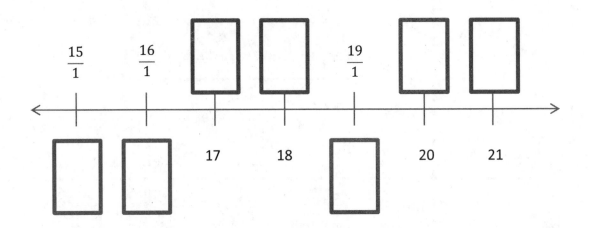

3. Explain the difference between these fractions with words and pictures.

$$\frac{5}{1} \qquad \frac{5}{5}$$

Lesson 25:   Express whole number fractions on the number line when the unit interval is 1.   EUREKA MATH

3 wholes

Lesson 25: Express whole number fractions on the number line when the unit interval is 1.

©2015 Great Minds. eureka-math.org
G3-M5-SE-B3-1.3.1-01.2016

109

This page intentionally left blank

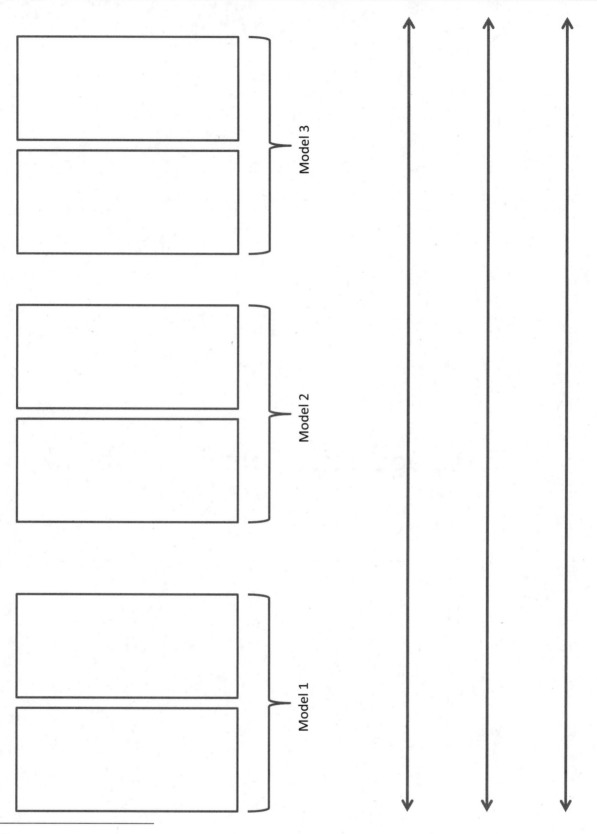

Model 3

Model 2

Model 1

6 wholes

Lesson 25:   Express whole number fractions on the number line when the unit interval is 1.

111

This page intentionally left blank

Name _____   Date _____

1. Partition the number line to show the fractional units. Then, draw number bonds using copies of 1 whole for the circled whole numbers.

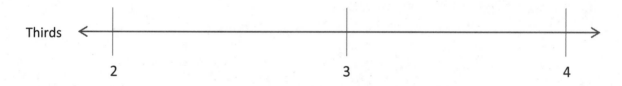

Halves

    0   1   2

0 = _____ halves   1 = _____ halves   2 = _____ halves

$0 = \dfrac{\square}{2}$ $1 = \dfrac{\square}{2}$ $2 = \dfrac{4}{2}$

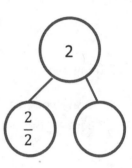

Thirds

    2   3   4

2 = _____ thirds   3 = _____ thirds   4 = _____ thirds

 $2 = \dfrac{\square}{3}$ $3 = \dfrac{\square}{3}$ $4 = \dfrac{\square}{3}$

2. Write the fractions that name the whole numbers for each fractional unit. The first one has been done.

	2	3	4
Halves	$\frac{4}{2}$	$\frac{6}{2}$	$\frac{8}{2}$
Thirds			
Fourths			
Sixths			

3. Sammy uses $\frac{1}{4}$ meter of wire each day to make things.

    a. Draw a number line to represent 1 meter of wire. Partition the number line to represent how much Sammy uses each day. How many days does the wire last?

    b. How many days will 3 meters of wire last?

4. Cindy feeds her dog $\frac{1}{3}$ pound of food each day.

    a. Draw a number line to represent 1 pound of food. Partition the number line to represent how much food she uses each day.

    b. Draw another number line to represent 4 pounds of food. After 3 days, how many pounds of food has she given her dog?

    c. After 6 days, how many pounds of food has she given her dog?

Lesson 26:   Decompose whole number fractions greater than 1 using whole
number equivalence with various models.

©2015 Great Minds. eureka-math.org
G3-M5-SE-B3-1.3.1-01.2016

EUREKA
MATH

Name _____   Date _____

1. Partition the number line to show the fractional units. Then, draw number bonds with copies of 1 whole for the circled whole numbers.

Sixths ⟵————————————————————————⟶

0   1   2

0 = _____ sixths   1 = _____ sixths   2 = _____ sixths

$0 = \dfrac{\square}{6}$ $1 = \dfrac{\square}{6}$ $2 = \dfrac{12}{6}$

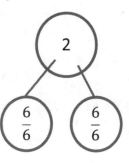

Fifths ⟵————————————————————————⟶

2   3   4

2 = _____ fifths   3 = _____ fifths   4 = _____ fifths

$2 = \dfrac{\square}{5}$ $3 = \dfrac{\square}{5}$ $4 = \dfrac{\square}{5}$

(2)   (3)

EUREKA MATH™

Lesson 26:   Decompose whole number fractions greater than 1 using whole number equivalence with various models.

115

©2015 Great Minds. eureka-math.org
G3-M5-SE-B3-1.3.1-01.2016

2. Write the fractions that name the whole numbers for each fractional unit. The first one has been done for you.

Thirds	$\dfrac{6}{3}$	$\dfrac{9}{3}$	$\dfrac{12}{3}$
Sevenths			
Eighths			
Tenths			

3. Rider dribbles the ball down $\frac{1}{3}$ of the basketball court on the first day of practice. Each day after that, he dribbles $\frac{1}{3}$ of the way more than he did the day before. Draw a number line to represent the court. Partition the number line to represent how far Rider dribbles on Day 1, Day 2, and Day 3 of practice. What fraction of the way does he dribble on Day 3?

Decompose whole number fractions greater than 1 using whole
                  number equivalence with various models.

EUREKA
MATH™

Name _____   Date _____

1. Use the pictures to model equivalent fractions. Fill in the blanks, and answer the questions.

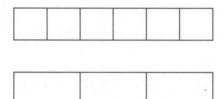

4 sixths is equal to _____ thirds.

$$\frac{4}{6} = \frac{\square}{3}$$

The whole stays the same.

What happened to the size of the equal parts when there were fewer equal parts?

What happened to the number of equal parts when the equal parts became larger?

1 half is equal to _____ eighths.

$$\frac{1}{2} = \frac{\square}{8}$$

The whole stays the same.

What happened to the size of the equal parts when there were more equal parts?

What happened to the number of equal parts when the equal parts became smaller?

2. 6 friends want to share 3 chocolate bars that are all the same size, which are represented by the 3 rectangles below. When the bars are unwrapped, the friends notice that the first chocolate bar is cut into 2 equal parts, the second is cut into 4 equal parts, and the third is cut into 6 equal parts. How can the 6 friends share the chocolate bars equally without breaking any of the pieces?

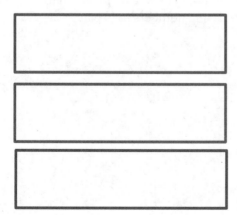

3. When the whole is the same, why does it take 6 copies of 1 eighth to equal 3 copies of 1 fourth? Draw a model to support your answer.

4. When the whole is the same, how many sixths does it take to equal 1 third? Draw a model to support your answer.

5. You have a magic wand that doubles the number of equal parts but keeps the whole the same size. Use your magic wand. In the space below, draw to show what happens to a rectangle that is partitioned in fourths after you tap it with your wand. Use words and numbers to explain what happened.

Lesson 27:   Explain equivalence by manipulating units and reasoning about their
                          size.

©2015 Great Minds. eureka-math.org
G3-M5-SE-B3-1.3.1-01.2016

EUREKA
MATH™

Name _____   Date _____

1. Use the pictures to model equivalent fractions. Fill in the blanks, and answer the questions.

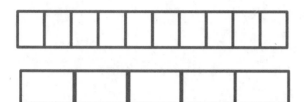

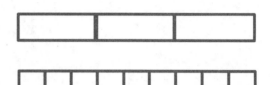

2 tenths is equal to _____ fifths.

$$\frac{2}{10} = \frac{}{5}$$

The whole stays the same.

What happened to the size of the equal parts when there were fewer equal parts?

1 third is equal to _____ ninths.

$$\frac{1}{3} = \frac{}{9}$$

The whole stays the same.

What happened to the size of the equal parts when there were more equal parts?

2. 8 students share 2 pizzas that are the same size, which are represented by the 2 circles below. They notice that the first pizza is cut into 4 equal slices, and the second is cut into 8 equal slices. How can the 8 students share the pizzas equally without cutting any of the pieces?

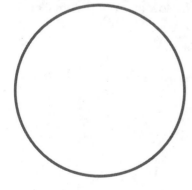
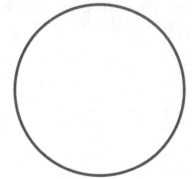

3. When the whole is the same, why does it take 4 copies of 1 tenth to equal 2 copies of 1 fifth? Draw a model to support your answer.

4. When the whole is the same, how many eighths does it take to equal 1 fourth? Draw a model to support your answer.

5. Mr. Pham cuts a cake into 8 equal slices. Then, he cuts every slice in half. How many of the smaller slices does he have? Use words and numbers to explain your answer.

Lesson 27:   Explain equivalence by manipulating units and reasoning about their size.

EUREKA MATH

Name _____   Date _____

Shade the models to compare the fractions. Circle the larger fraction for each problem.

1. 2 fifths

    2 thirds

2. 2 tenths

    2 eighths

3. 3 fourths

    3 eighths

4. 4 eighths

    4 sixths

5. 3 thirds

    3 sixths

6. After softball, Leslie and Kelly each buy a half-liter bottle of water. Leslie drinks 3 fourths of her water. Kelly drinks 3 fifths of her water. Who drinks the least amount of water? Draw a picture to support your answer.

7. Becky and Malory get matching piggy banks. Becky fills $\frac{2}{3}$ of her piggy bank with pennies. Malory fills $\frac{2}{4}$ of her piggy bank with pennies. Whose piggy bank has more pennies? Draw a picture to support your answer.

8. Heidi lines up her dolls in order from shortest to tallest. Doll A is $\frac{2}{4}$ foot tall, Doll B is $\frac{2}{6}$ foot tall, and Doll C is $\frac{2}{3}$ foot tall. Compare the heights of the dolls to show how Heidi puts them in order. Draw a picture to support your answer.

Lesson 28: Compare fractions with the same numerator pictorially.

EUREKA MATH™

©2015 Great Minds. eureka-math.org
G3-M5-SE-B3-1.3.1-01.2016

Name _____   Date _____

Shade the models to compare the fractions. Circle the larger fraction for each problem.

1.   1 half

     1 fifth

2.   2 sevenths

     2 fourths

3.   4 fifths

     4 ninths

4.   5 sevenths

     5 tenths

5.   4 sixths

     4 fourths

6. Saleem and Edwin use inch rulers to measure the lengths of their caterpillars. Saleem's caterpillar measures 3 fourths of an inch. Edwin's caterpillar measures 3 eighths of an inch. Whose caterpillar is longer? Draw a picture to support your answer.

7. Lily and Jasmine each bake the same-sized chocolate cake. Lily puts $\frac{5}{10}$ of a cup of sugar into her cake. Jasmine puts $\frac{5}{6}$ of a cup of sugar into her cake. Who uses less sugar? Draw a picture to support your answer.

Lesson 28:   Compare fractions with the same numerator pictorially.

EUREKA
MATH™

Name _____ Date _____

Label each shaded fraction. Use >, <, or = to compare. The first one has been done for you.

1.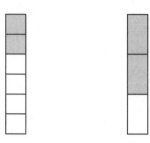

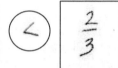 $\dfrac{2}{6}$ $\left(<\right)$ $\dfrac{2}{3}$

2.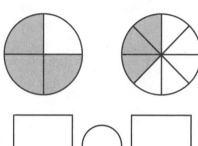

3.

4.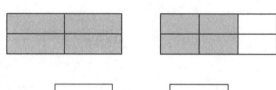

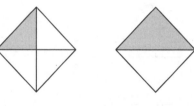

5. Partition each number line into the units labeled on the left. Then, use the number lines to compare the fractions.

halves

0   1

fourths

0   1

eighths

0   1

a. $\dfrac{3}{8}$ ◯ $\dfrac{3}{4}$

b. $\dfrac{4}{4}$ ◯ $\dfrac{4}{8}$

c. $\dfrac{2}{4}$ ◯ $\dfrac{2}{8}$

EUREKA MATH

Lesson 29: Compare fractions with the same numerator using <, >, or =, and use a model to reason about their size.

125

©2015 Great Minds. eureka-math.org
G3-M5-SE-B3-1.3.1-01.2016

Draw your own model to compare the following fractions.

6. $\dfrac{3}{10}$ ◯ $\dfrac{3}{5}$ 7. $\dfrac{2}{6}$ ◯ $\dfrac{2}{8}$

8. John ran 2 thirds of a kilometer after school. Nicholas ran 2 fifths of a kilometer after school. Who ran the shorter distance? Use the model below to support your answer. Be sure to label 1 whole as 1 kilometer.

9. Erica ate 2 ninths of a licorice stick. Robbie ate 2 fifths of an identical licorice stick. Who ate more? Use the model below to support your answer.

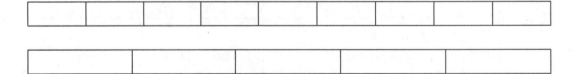

Lesson 29:   Compare fractions with the same numerator using <, >, or =, and use a
                              model to reason about their size.

EUREKA
MATH™

Name _____   Date _____

Label each shaded fraction. Use >, <, or = to compare.

1.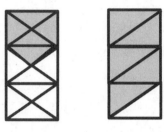

2.

3.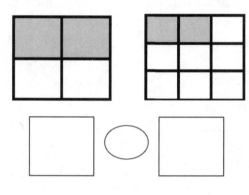

4.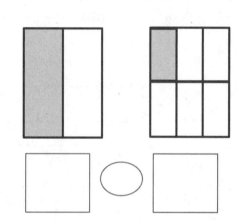

5. Partition each number line into the units labeled on the left. Then, use the number lines to compare the fractions.

thirds   0   1

sixths   0   1

ninths   0   1

a. $\frac{2}{6} \bigcirc \frac{2}{3}$ b. $\frac{5}{9} \bigcirc \frac{5}{6}$ c. $\frac{3}{3} \bigcirc \frac{3}{9}$

Draw your own models to compare the following fractions.

6. $\dfrac{7}{10}$ ◯ $\dfrac{7}{8}$ 7. $\dfrac{4}{6}$ ◯ $\dfrac{4}{9}$

8. For an art project, Michello used $\dfrac{3}{4}$ of a glue stick. Yamin used $\dfrac{3}{6}$ of an identical glue stick. Who used more of the glue stick? Use the model below to support your answer. Be sure to label 1 whole as 1 glue stick.

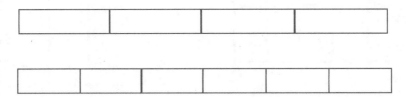

9. After gym class, Jahsir drank 2 eighths of a bottle of water. Jade drank 2 fifths of an identical bottle of water. Who drank less water? Use the model below to support your answer.

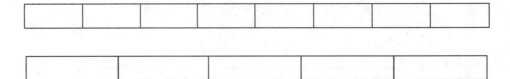

Lesson 29:   Compare fractions with the same numerator using <, >, or =, and use a
                       model to reason about their size.

EUREKA
MATH™

Name _____ Date _____

Describe step by step the experience you had of partitioning a length into equal units by simply using a piece of notebook paper and a straight edge. Illustrate the process.

Lesson 30:   Partition various wholes precisely into equal parts using a number
              method.

129

©2015 Great Minds. eureka-math.org
G3-M5-SE-B3-1.3.1-01.2016

This page intentionally left blank

lined paper

Lesson 30:   Partition various wholes precisely into equal parts using a number
             method.

131

©2015 Great Minds. eureka-math.org
G3-M5-SE-B3-1.3.1-01.2016

This page intentionally left blank